NEVER
the HOPE
ITSELF

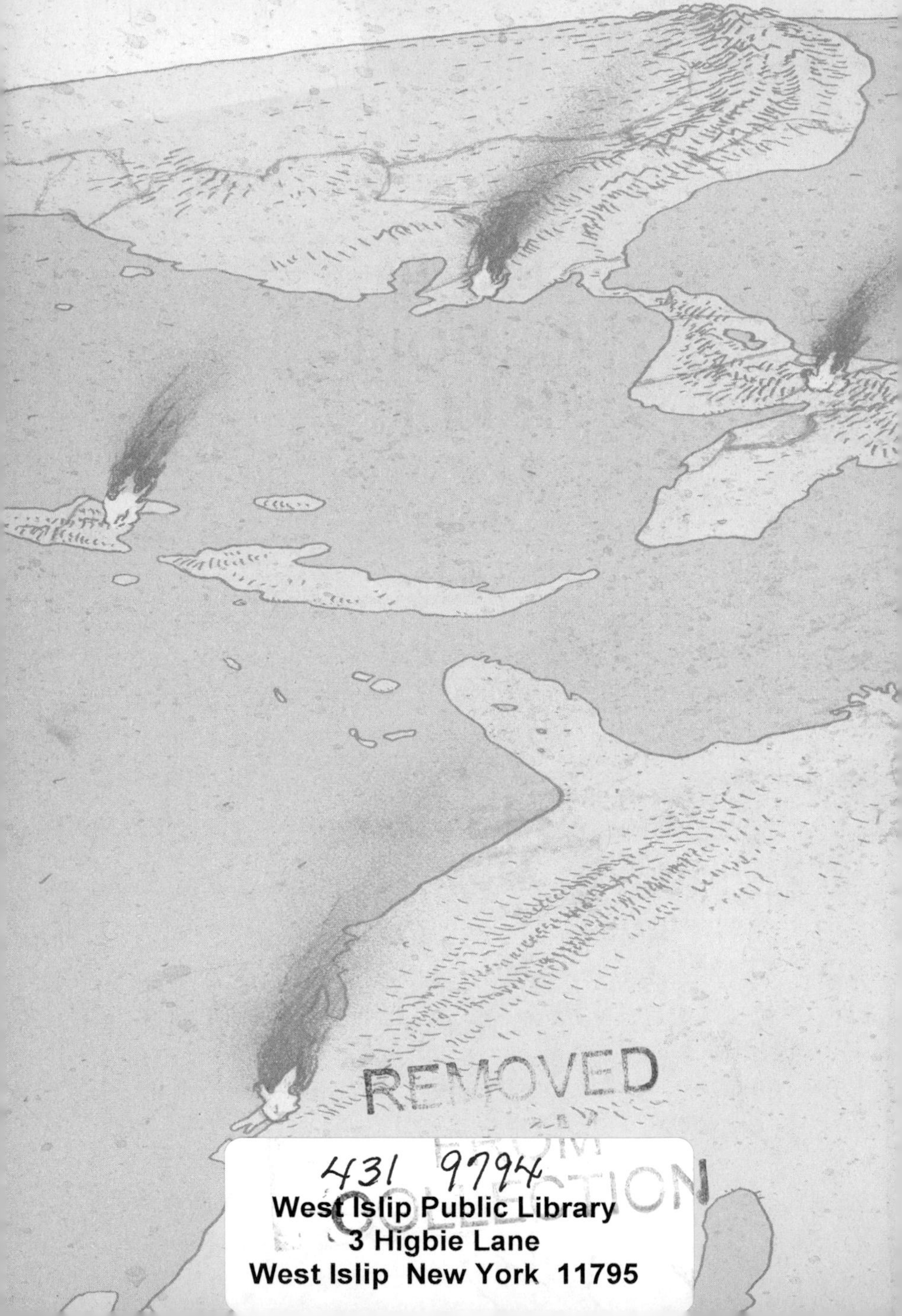

NEVER the HOPE ITSELF

LOVE AND
GHOSTS IN
LATIN AMERICA
AND HAITI

GERRY HADDEN

HARPER ◗ PERENNIAL

NEW YORK • LONDON • TORONTO • SYDNEY • NEW DELHI • AUCKLAND

HARPER ● PERENNIAL

NEVER THE HOPE ITSELF. Copyright © 2011 by Gerry Hadden.
All rights reserved. Printed in the United States of America.
No part of this book may be used or reproduced in any manner
whatsoever without written permission except in the case of
brief quotations embodied in critical articles and reviews. For
information address HarperCollins Publishers, 10 East 53rd
Street, New York, NY 10022.

HarperCollins books may be purchased for educational,
business, or sales promotional use. For information please write:
Special Markets Department, HarperCollins Publishers, 10 East
53rd Street, New York, NY 10022.

FIRST EDITION

Designed by John Lewis

Map on title page © Stephan Liechti

Library of Congress Cataloging-in-Publication Data is available
upon request.

ISBN 978-0-06-202007-9

11 12 13 14 15   OV/RRD   10 9 8 7 6 5 4 3 2 1

For Mouche and Lula and Nino and the boy

and to the crooks, cranks, saviors, and saints
of Latin America and the Caribbean

PART ONE

R on and I were hurtling down Haiti's National Highway 2 when his beat-up red Cherokee Blazer came to
a sudden, shuddering stop. My forehead bounced against
the front windshield. We both jumped out of the truck and
cursed. It was getting dark and I couldn't afford to be outside the capital any longer.

From the dusty ridge we could see the cruel and crowded
slums of Port-au-Prince on the horizon below, simmer-
ing like a coal fire against the blue-green sea. Haiti's NH2
was officially a highway, but it looked and felt more like a
dry riverbed. Potholes deep enough to entrap cars pocked
its surface. There were no guardrails between us and rocky
oblivion on most of the curves as we descended from Haiti's
Central Plateau toward the capital. On some stretches the
road narrowed to such a degree that traffic could move only
in one direction. Often that meant yielding to giant, swaying Korean-made cargo trucks overflowing with Haitian

laborers on their way to or from the fields and the few factories that were still operational in that country run nearly
into oblivion. But when Ron's Cherokee broke down that
late May afternoon we were alone on the mountainside.

After hours of jarring journey the truck's driveshaft had
finally broken, detached from the front axle, then embed-
ded itself like a jouster's lance into the stony road. Ron tried
hammering it free with a rock. The first one broke in two.

"Here," I said, handing him another. "A bigger one."

But it was useless. The weight of the truck held the driveshaft fast in the ground. Ron shimmied out from under the
truck and fished a small crowbar out of a toolbox and set to
work banging again. After several minutes we climbed back
into the vehicle.

"I'm going to call Triple A," I said. I knew the joke was bad
but I was nervous. I'd slipped out of Port-au-Prince just two
days after legislative elections and the results were still not
in. Things were tense in the capital; some fifteen people had
been murdered in pre-election violence. In the last week several small bombs had been set off, and everyone was braced
for more trouble. But when an Organization of American
States election official told a group of us reporters that the final tally would be delayed at least another few days—ballots
from the deep countryside were still arriving on the backs of
donkeys—I decided to take a chance and slip away to gather
tape for another story. This was my first assignment abroad
for National Public Radio and my pulse was ticking a few
beats per minute above the recommended rate for a man of

my age, thirty-three. The attacks of September 11 were still more than a year away and Americans and the American government were paying close attention to this tiny but potential tinderbox of a country. Our audience was huge and interested and I was the untested new guy.

Ron was walking in a big circle, holding his phone over his head. "No coverage," he said. "Damn it damn it damn it." Then he pocketed his phone and clenched his fists and said confidently, "To plan C!"

Plan C apparently consisted of going back to plan A because he slipped back under the truck and resumed his hammering. I stayed in the cabin replaying the sounds of a small sugar mill crushing freshly cut cane. The juicy squeak of the metal press, the sucking plod of the gray horse turning the mill in muddy circles. The laughter of children as sap ran and the man with the microphone winked. The mill was at the heart of a short piece on rural cooperatives that I hoped would shed a more positive light on Haitians and their efforts to lift themselves out of poverty. Nearly every Haitian I had met so far had implored me to do a "positive" story about the country. We journalists were always reporting the bad news.

I had my headphones on but I could still hear the clang of Ron's crowbar against the driveshaft like the distant call to supper on some cowboy ranch. Then I heard what I thought was a sort of yelp. The truck lurched forward and began to roll down the steep incline.

Through the open driver's side door I saw Ron rolling out from under the Cherokee, kicking up a cloud of white dust.

Apparently he'd freed the driveshaft. But he'd forgotten to secure the truck's emergency brake. I was sitting in my tangle of cables, microphones, and minidiscs and the emergency brake was on the far side of the cabin, low to the floor, too far for my foot to reach it in the short moment I had to react. As the truck picked up speed I chucked myself out my door. I dragged half my gear with me and landed prone on the gravelly ground. My headphones were still strung around my neck and the cable quickly went taut; it must have become caught on some part of the truck. Without thinking I grabbed fast to the headphones and was promptly rolled over. The cable snapped and boomeranged back in my face. I stood up in time to watch our wounded red vehicle bounding toward the next downhill curve in the road, now doing about twenty miles an hour and gaining speed, its perverse-looking driveshaft scraping along the road. When it reached the curve I bid a hasty good-bye to a week's worth of recordings still scattered across the floor of the truck's cabin. But then the Cherokee struck a huge rock on the edge of the precipice and its front end shot up in the air. The steering wheel spun hard toward the mountain, and when the wheels touched down again the truck lurched back across the road and smashed into the rocky mountainside, coming to a stop on two wheels. Ron and I ran down the hill.

"Fuck me," Ron said, "that was lucky."

"You got any cigarettes left," I said. I didn't smoke. I quickly gathered my gear from the floor of the truck. We walked over to the big boulder that had saved Ron's truck and probably us as well and sat down on it. The sun was dropping, the gray stain of the city in the distance darkening now. "If I ever go to grad school I'm going to study cable management," I said, untangling the wad of wires at my feet. Ron had grown morose.

"I guess we're walking home," I said.

"Are you kidding?" Ron said, sitting up straight and taking a drag on his cigarette. In his white undershirt and with white dust covering his face and mustache he looked like a baker on a smoke break. "We can't walk down this mountain. After dark it's filled with bandits."

"Bandits?"

"We've got to get the truck moving. If we leave it behind they'll strip it and torch it. "

"And us?"

Ron ignored the question. The truck's prognosis had worsened. In the crash the rear axle had slipped out of alignment a couple of inches. We managed to get the driveshaft back in position, but we couldn't secure it. Ron started his hammering again. It was the only sound on the entire mountainside—clang, clang, clang, clang. Then some young Haitian men came walking around the bend in the highway, shirtless and bony, and helped us push the truck more or less back onto the road. One of them, a severely bucktoothed kid with long lanky limbs, had an idea: we should start the truck, drop it in reverse, and ram it tail-end into a boulder thus forcing the rear axle forward and the driveshaft back into its socket.

"Why not?" Ron said.

Six of us pushed a big rock into the road behind the truck and Ron slammed the Cherokee against it in reverse. It cracked the bumper but the rear axle shifted forward again and the driveshaft jammed back in place, at least temporarily. When Ron saw that it had worked he smiled at the Haitian kid's ingenuity. He said something to the boys in Kreyol and tipped them some *gourdes* for their trouble. The boys skipped and disappeared up the road. We were at least an hour from any village, and I wondered where on earth they'd come from and where they were going.

Ron dropped one last time beneath the truck to secure his broken driveshaft, wrapping the joint in some wire, and we started slowly down the mountain again just as the sun went down. An hour later we limped into Port-au-Prince doing five miles an hour, the truck rolling diagonally forward in a mechanical canter. We trundled past the crowded market stalls and the public water taps with their long lines of people waiting patiently or not so with big plastic drums. Past the narrow, tin-roofed shops and the families milling along the dirt lanes that perhaps in better days had sidewalks but now did not. There were no streetlights and the air was thick with the smell of charcoal fires. Clouds of cooking spice and exhaust inseparable. I had the sensation that we were at once in the city and somewhere very removed from it. There were no signs of election-related trouble, and Ron told me that the radio was saying as much.

But I had arrived back just in time. The election results were announced the next day, giving eighteen out of nineteen open senate seats to former president Jean-Bertrand Aristide's Fanmi Lavalas party. Fanmi Lavalas's sweep seemed a clear indication that Aristide would win the presidency again in November. Under Haiti's constitution a president can serve two terms. What we didn't yet know was that officials had counted the votes using a faulty method that unfairly favored Aristide's candidates. Nor that the ensuing controversy would set off yet another degenerating cycle of violence and political stagnation in a country that has known only brief moments of real freedom. About the only thing I knew at the time was that I'd managed to get myself very far from home in a very short span of time.

Just a week earlier I was living in Seattle, Washington, mentally preparing myself to be fitted for the robes of a Buddhist monk. Now this would-be monk was a foreign correspondent. My life had changed so suddenly that I'd barely had time to pack my bags, much less reflect on the track switch my karma had just thrown. But as I rode along in Ron Bluntschli's ailing truck through Port-au-Prince, clutching my radio gear tightly to my lap, I was thinking, These journalist's clothes fit pretty well too.

I'd been at this job for less than a week. NPR had hired me for the position as I was finishing my fifth year of local reporting for public radio station KPLU in the Emerald City.

I was fluent in Spanish and NPR's foreign desk chief, Loren Jenkins, liked the fact that, outside of work, I played Latin music in various clubs around town. The next thing I knew I was on a plane to D.C. for training at NPR's swank Massachusetts Avenue headquarters. Jenkins introduced me to the news staff.

"Listen up, everyone, he's fluent in Spanish." A few heads nodded approvingly.

"He plays Cuban drums." A lot more heads moved.

Jenkins then passed me off to the sound engineers. A bearded man in a conservative tie handed me a digital recording device, a couple of microphones, a bag full of cables and wires, and a hard plastic suitcase that weighed about forty pounds.

"That's your satellite telephone," he said. "Whatever you do, don't stand in front of the dish."

"Because?"

"Because it'll soften you up like ramen noodles. It emits microwaves."

We went up on the roof, and he had me fiddle with a compass and a map until I had the position of our satellite more or less pinpointed. I aimed the dish, moved behind it, and dialed. The engineer's cell phone rang.

"You got it," he said, rejecting the call and pocketing his phone. "Now you're a pro."

Before leaving D.C. I asked Jenkins if I was going to have a chance to meet Philip Davis, the reporter I was replacing.

"It'd be nice to get the lay of the land from him. Contacts, tips, the do's and don'ts of living in and working out of Mexico City."

"Philip's long gone," Jenkins said. "Married a Mexican girl. He's back on the national desk, in Miami."

"How will I get the keys to NPR's bureau in Mexico City?"

"There is no such thing to get," Jenkins said. "You find a place when you get down there."

It didn't really sink in until I was boarding my flight to Mexico City: I was flying off to start what was, in my dreams at least, my dream job. I felt like telling the flight attendants. I was startled by how easily I'd set aside another equally important project of mine: my meditation retreat. My three-year meditation retreat, for which I'd been practicing for as many years. I'd been slated to enter a retreat center in New South Wales, Australia, in a few months' time. My teacher had given me the green light at a recent teaching in northern California.

"I have the time, the inclination, and the money," I'd told the saffron-robed Dzongsar Khyentse Rinpoche as I bowed before him. Hundreds of students were waiting behind me, in line to pay their respects as well. "Will you accept me at the *gompa*?"

"Yes," he said, smiling. "Why not?"

His answer nearly sent me levitating. I'd finally made it. I was ecstatic. There was no backing out now. You didn't just ask the spiritual leader of Bhutan for retreat permission lightly. And you didn't squander a rare yes. A year earlier, when I'd made the same request, Rinpoche had responded, "You cannot escape from work."

So I'd gone back to work. But now, happily, things had changed. I floated back to my seat in the audience across a sea of burgundy-colored meditation mats and piles of prayer beads. I wondered what change Rinpoche had seen in me. Now I believe that he hadn't seen any change at all. I think maybe what he saw was that I would never make that retreat.

As my plane took off from D.C. I said adios to my search for meaning via the sedentary observation of my mind. Instead, I thought, I'm going to have my mind blown away. I thought of the hippie in the film version of the musical *Hair* who accidentally gets shipped off to the Vietnam War. But my flight didn't feel tragic, or even like a mistake. Enlightenment had been one of my goals, but working for NPR was another. I was in heaven, to borrow from the theists.

During the flight I had time to look back on it all. On how I'd gotten here, on how I'd become a Buddhist. I was raised Presbyterian but nearly from the beginning my soul, that insoluble pillar of theistic reasoning, had been under assault. I think I first realized it one fall day when I was seven years old. My sister, Hartley, two years older than me, came into my room and unwittingly delivered the existential concussion.

"Wanna see your real name?" she asked.

I got up and padded after her down the stairs. All I knew about my biological past was what my adoptive parents had told me: that my birth father was Swedish. From there, a thousand fantasies already. My sister, who was adopted at birth as well, apparently had English ancestors. Now she was leading me quietly into our father's study. The off-limits zone. I don't know if he was home, but I was nervous. So was Hartley. She lifted up a dusty bowling trophy on a shelf and slid a small silver-colored key out from under its base.

"Where did you find that?" I said.

"Duh. Where do you think?"

She crossed to the metal filing cabinet and worked the lock easily; so she'd been here before. I was shocked. I was supposed to be the mischief maker in the family. My sister pulled out two legal-size manila folders. She handed me one.

"Here you go. Lars."

"Lars?"

She took the folder back, opened it.

"Here." She pointed to the bottom of a sheet of paper with lots of things crossed out in black ink. But the first and last name typed near the bottom had not been blacked out—as, I would later learn, they should have been, according to New York State adoption laws.

Hartley showed me my birth name, Lars Besser, and hers. Then she quickly filed away the folders and locked the cabinet and set the key as she'd found it. I went back to my room. All of three minutes had passed. I had been doodling great white sharks on loose-leaf paper, and I went back to it. But I was no longer a little boy. I was now two little boys. One was named Gerald, after his adoptive grandfather, and the other, Lars. The son of a Swede. Background blacked out. My sense of identity, already resting on a shaky platform, had just been toppled by an oceanic blowout.

Later that same year, just as I was getting used to being two people, I became less than one. For a long time after, I blamed the Muppets. I'd been playing "Swedish Chef" with some friends in the neighborhood, inserting fallen dogwood leaves into the spinning blades of a rusty old push lawn-mower. In a fit of enthusiasm I inserted one leaf too far. The whirring blades clipped off the top joint of my left middle finger. When the ordeal of bleeding and suturing had ended and my eyes had adjusted to the light in my hospital room, I asked my mother the question that was most bothering me.

"Mom," I said, "Where is my finger?"

"I'm afraid they couldn't reattach it," she said sadly.

This upset her greatly because she was imagining there were things I would now never be able to do: play concert cello, operate on brains—anything requiring a high level of dexterity and ten birth fingers. I could have cared less. At seven, these were not issues for me.

"So where is it?" I said.

"They would have thrown it away," she said. "They have a special garbage. An incinerator."

"What is that?"

"A big fire."

I could see it all clearly then. The pale, lacerated stub.

The fingernail catching fire first. Roasting in a huge outdoor campfire ringed by stones. Men and women in white lab coats standing around making sure the flames consumed it fully. I could also imagine how it must have hurt. I'd burned my fingers just recently, setting tiny plastic army soldiers on fire in a pal's garage. And this got me started on a conundrum. I certainly hadn't felt any pain from the hospital incinerator. So my fingertip couldn't have been Me. Me was still in bed dealing with the pain on the other side of the wound. Fair enough. But what, I wondered, would have happened if I'd cut off my entire hand? Where would Me be then? Not in the hand, I thought. Suppose it had been my whole arm? Or, what if every year those same docs in lab coats cut off a little chunk of me until less than half of my original body was left? Then I'd be a Me in the minority, as it were. Might my soul have escaped by then? Where does it reside and how much room does it need anyway? Does it even exist?

Doppelganged, then cloven, I carried these confusions with me into adulthood. As the years passed the identities of the people around me seemed to become more and more solidified. College majors, careers, political views, brands of beer, brands of shoes, tolerances and intolerances. My own identity remained mushy. I had a driver's license, a diploma, a history verifiable in the dozens of thick photo albums on a shelf in our living room, the innumerable witnesses to my life, friends, girlfriends, the mirror in the bathroom. Yet everything I was, all of my experiences, ultimately hinged on the sometimes terrifying randomness of my adoption. Who would I have been if I hadn't been adopted? Or if I'd been taken in by a different family? By carnies? Or physicists? Or zillionaires? As time passed I asked myself if my listing platform might ever be righted. And then I stumbled upon a philosophy that said, Wrong question. The only question that matters is, Who's asking the question?

Look, the Buddhist masters were saying, and you won't ever find anyone. Not anyone with any sort of permanent, free-standing identity at least. No permanent self. There was this giant misunderstanding. They called it "ego." And it didn't really exist. There was nothing to defend after all. The trick was to watch that ego carefully, to watch it until it deconstructed itself under the light-handed scrutiny of meditation.

What a relief! This low-level but ever-present mental anguish might dissolve under investigation. It might give way to a fuller experience of life. To the realization that everything is fundamentally as it should be. It might mean the end of fear.

When I first read this, at age twenty-eight, I rejoiced. For the first time, instead of feeling out of step with the world I felt one step ahead of it. I could already imagine myself a monk.

I came back to earth, along with my plane, at Mexico's Benito Juarez International Airport. I booked myself straight into a downtown hotel and hit the streets looking for a place that could serve as the NPR bureau and my home. I set out for the trendy and supposedly safe Condesa neighborhood on the advice of several sources.

As the newest arrival on NPR's foreign desk I felt tremendous pressure to prove myself. It seemed like just yesterday that I'd filed my first NPR spot, a one-minute news story, for the top-of-the-hour newscast in Washington, D.C.

I'd been working quietly at the KPLU bureau when a bulletin came across announcing that police were evacuating downtown Seattle. I was alone in the office so I grabbed my recording kit and ran the few blocks to the scene. There, police were stringing up yellow tape.

"What's going on?" I asked one.

"Bomb scare," he said, pointing. About a hundred yards away, in the center of a pedestrian square, a graffiti-covered pickup truck stood abandoned, its tires punctured. Propped up in the bed of the truck I could see a six-foot-tall sculpture of what was clearly a human heart. I ducked into a pay phone and called my editor, Erin Hennessey, who was down at the station's broadcast center in Tacoma.

"There's a truck out here with a heart in it."

"NPR's already called and wants spots," she said. "So does NBC radio and a half dozen other news outlets. You want to become famous?"

"Yes, ma'am."

"You've got five minutes," she said. "Write something up and call NPR's news desk. They'll take you in."

I hung up and reached for my pen and notebook. But I'd

forgotten them in my haste to reach the scene. I ditched the phone booth, ducked under the police tape, and sprinted into a nearby Starbucks. Coffee addicts were waiting calmly in a long line for their fixes despite the excitement outside. I ran to the register and grabbed a handful of napkins.

"Excuse me!" I said out loud. "Can anyone lend me a pen?"

The caffeine-crazed patrons just looked at me.

"Does anyone have a pen? Please!" Then I saw one by the cash register. I snatched it up.

"That's mine," said a stoned-looking barista.

"I'll be right back!" I yelled and ran back out into the street. I nearly collided with a cop who scolded me and told me to get back behind the police line. I assured him that that was my intention, then ducked back into the phone booth. Sweat was dripping down my temples. It was difficult to write on the tiny napkins with their big green logos taking up most of the space, but I gave it my best shot: "A suspicious vehicle left on the Westlake Mall is being investigated by police . . ." I called NPR. The switchboard patched me into the newscast unit and a producer gave me a countdown. My hands shook as I read the spot from my napkin script. Some of the words had gotten wet and smeared.

When I was through there was silence. Then the producer came back on the line and said, "Nice job. Call us again with an update in half an hour." At 10:00 p.m. that night I was still filing stories, burning through my reel of recording tape in one of our two small recording studios.

I was up so late because the story had evolved over the

course of the day from a bomb scare into something decidedly less deadly and more intriguing. The pickup truck belonged not to terrorists but to a mentally unstable local artist named Jason Sprinkle. Sprinkle, known around town as a provocateur, had abandoned his vehicle with its odd cargo in a place he considered symbolic of America's rampant consumerism. He'd jumped out and popped the tires and run like hell. It was performance art, a smallish protest against shopping culture in an urban stronghold of sparkling outlets and chain stores.

What Sprinkle hadn't counted on was setting off a stampede of terrified office workers and shoppers. This was before 9/11, but the world had long grown skittish when it came to terrorist threats. Sprinkle was eventually caught and thrown in jail on terrorism-related charges. In the end he was held in the clink for a month, given a year's probation, then freed. But friends said Sprinkle's stay in jail broke his already fragile spirit. Several years later, while I was living in Mexico, I read that he'd been struck and killed by a freight train in Mississippi. It was unclear whether he'd taken his own life. I took the occasion to thank him again. If it hadn't been for that forlorn and grotesque rendering of a heart—his own maybe, made manifest—abandoned in the shadow of Abercrombie and Fitch, I might never have ended up on a plane to Mexico.

On my first day in Mexico City I cruised the Condesa neighborhood's tree-lined streets searching for FOR RENT signs on

doors and windows. I strolled past wide three-story houses
stuccoed in royal blue, bright orange, salmon. The sidewalks
buckled block after block, uprooted by flowering magnolias.
A man rode by on a bicycle towing an aluminum oven filled
with steaming tamales, the tinny blare of his recorded sales
pitch cycling again and again from an old loudspeaker. *Ricos
tamales oaxaqueños . . . Ricos tamales oaxaqueños . . .* An Indian sharpened scissors in a doorway under the absent gaze
of somebody's grandmother. Schoolgirls with black pony-
tails and Hello Kitty knapsacks walked by. Two squat men
in blue uniforms guarded a hardware store with shotguns.
The Condesa was intriguing at first pass, but house hunting
was not how I wanted to be spending my first days on the job.

Haiti saved me. I believed then that few people had ever
said such a thing, but now I know better. That night in Mexico City my hotel phone rang. It was my new editor, Paul
Glickman.

"I need you on a plane to Port-au-Prince ASAP," he said.
"We want you there for the legislative elections. Call me as
soon as you get to your hotel." Glickman was a former reporter who'd covered the cold war proxy conflicts in Central
America during the 1980s. When I was working on a story
with him over the phone he had a habit of pausing for so
long I'd start to wonder whether he was thinking or drinking a coffee down at the corner café. He was always thinking.
Glickman was thorough and he knew his stuff. Editing with

him each day was like playing tennis with someone better than you. He improved your game.

"You got it," I said. "I'll call you from my hotel."

Hotel? I had no idea where to stay, or even how to find out where to stay. I logged online with my NPR laptop and hastily booked a place over the Internet that looked promising: a spiffy, whitewashed Holiday Inn just a stone's throw from Haiti's presidential palace. Then I scrolled through the PalmPilot addresses I'd inherited from Philip Davis until I found the name of an American woman named Karla Bluntschli in Port-au-Prince. I called her then to see if she could work as my driver and translator, or fixer, while on the ground. Happily, she agreed.

The following afternoon I was on the last plane from Miami to the tiny Caribbean nation that shares the island of Hispaniola with the Dominican Republic. I was lucky I made it. Airlines were canceling flights because of a spate of violence in Port-au-Prince that was threatening to derail the elections. Small homemade bombs had been exploding around the capital: one at the entrance to the CEP, Haiti's electoral council; one at an outdoor market—a man's finger was blown off; another just outside the airport. Some fifteen people had been murdered in electoral violence over the last several weeks. As we were pushing and shoving our way off the plane—Haitians always mobbed each other with the same vigor whether they were returning to their country or leaving it—a chubby American stewardess grabbed me by the arm and said in a soft southern accent, "Excuse me, sir, are you American?" I nodded yes as I struggled to keep my balance. She squeezed my elbow in a motherly gesture and with worried eyes said, "*Puh-leeze* be careful."

That woman, with her coiffed blond hair and her American Airlines uniform, had probably never ventured past Haitian customs even in the best of times. Her concern spooked the hell out of me.

Walking out of the airport I was swarmed by young men who tried to take my bags from me. Some spoke English, some didn't; all of them were yelling. I'd arranged for Karla to pick me up, and soon I spotted her waiting for me in the parking lot—she was the only white woman around and she was craning her neck in the way people do when searching for others. Karla had big tired eyes and a graying ponytail and she strode straight over when she saw me. In near perfect Kreyol she got the crowd of men to back away from me. I didn't know what she was saying but her good humor set me at ease.

"What was that all about?"

"It was about a country where there are no jobs," she said. "These guys are just trying to make a buck. Get in!"

We piled into her beat-up red Cherokee and pulled out of the parking area. It was muggy and dusty. The main road to town was unpaved and scarred with huge potholes like those I encountered later that week on National Highway 2, where I nearly got stranded with Karla's husband, Ron.

In and among the slow-moving cars surged crowds of people, the descendants of slaves from various West African nations whom the French had thrown together in the six-teenth century in one of the cruelest chapters of the colonial period.

As Karla's truck jerked and jolted along the road she glanced at me to see if I was holding on okay. "If a Haitian president could just come to power and stay in power long enough to pave Port-au-Prince's roads," she said, "he would be loved for eternity." She and Ron had come to Haiti some fifteen years earlier as Christian missionaries, out to "save" the Haitians. But unlike most of the earnest, naïve do-gooders, Karla told me, she and Ron quickly saw the futility of trying to impose non-Haitian ideas on Haitians.

"You wanna know what happens?" she said as we sat in traffic, "I'll tell ya. Lots of well-meaning, helpful folk come to Haiti to save it. They set up a clinic or micro-lending in-stitution or farming cooperative, fund it, teach the locals how to manage it, oversee the project for a few years, and then leave. And as soon as they're gone their project falls apart—or better put, the Haitians take it apart—salvaging whatever is of interest to them and generally going back to how they've always done things, for better or worse."

"Why?"

"Haitians are still deeply scarred by the colonial experi-ence. The only way to survive the French was through se-crecy and duplicity. Haitians are still extremely cautious with outsiders. And like anybody, they don't like being told how to do things."

"But you all stayed."

"We're still here."

Not only had Ron and Karla stayed on—they'd even broken with their American church when their newfound respect and understanding of Haitian religious traditions put them at odds with their own spiritual leaders.

"You can imagine the conversation," she laughed. "Me and Ron explaining to our minister that Voodoo is as valid a practice as any. Didn't go over well."

We arrived at a medium-sized house perched on a hillside in the Delmas neighborhood of Port-au-Prince. The Bluntschlis were renting this place while slowly building their dream house on a mountainside outside the capital. There, among other projects, they were designing an interactive outdoor "freedom museum" in a neatly tended field. They hoped to have it finished by 2004, when Haiti would mark the bicentennial of its bloody independence from France and the subsequent two centuries of shafting by the so-called international community scared pale by the world's first successful slave rebellion.

After a brief lunch and some strategizing for the week Karla dropped me off downtown at a wretched hotel along the Champs de Mars, not far from the stately, whitewashed National Palace where farmer-turned-president René Préval was happily approaching the end of his term with his skin still intact. I looked around for the Holiday Inn I'd seen online. After a brief study of my notes and the semi-ruined dump bearing the street number of my virtual booking, I confirmed that it was in fact the same building. But it was clearly no longer part of the American chain.

"Hasn't been for years," Karla laughed as she drove away. "See you in the morning!"

Nor apparently had anyone been keeping it up since the chain dropped it. If I'd had a gallon of paint I might have slapped it on the façade myself. After unpacking my bags in my room I sat down on the old cast-iron bed and took several deep breaths. I let my mind clear for a moment. Then I gathered up my gear and headed for the street. I had little idea where to start, how to shape the stories that would preview the upcoming elections, in part because I hadn't yet called Glickman. First I wanted to get some small feel for the place. As I crossed the long open interior patio of the hotel I had the sense that I was the only guest there. The staff, from the cooks to the pool cleaner, were standing by the empty pool bar watching me. The only sound was the rubber squeak of my boots against the smooth concrete floor. I pushed open the front door.

Outside along the main street cars coughed gray smoke and colorfully painted pickup-truck taxis drove past and crowds of people picked their way between them. About a dozen desperately poor young men accosted me. One guy, a mulatto with frizzy black hair and a filthy Jimi Hendrix T-shirt, introduced himself in English as Robinson.

"I was deported from Florida last year, man," he began without prompting. "I been in the States for twelve years, since I was seven years old. I don't know what the hell I'm doing here."

"If it makes you feel any better," I said, "that makes two of us."

"I've had to learn Kreyol all over again."

"Why were you deported?"

"I never had papers, man. My parents never had papers. One day the police caught me. A few days later I was on a plane."

The abruptness of Robinson's displacement made me feel better in some strange way. He took me by the arm and led me across the street. I was recording now.

"This is where I work," he said. We were standing next to an open manhole. Robinson grabbed a rope and lowered a bucket several feet down and scooped up a load of filthy brown water into which he dipped an old shirt. Wringing the shirt with great effort he then began "cleaning" the cars parked along the road. What he was actually doing was swirling muddy water on the vehicles. But it was something.

"You can see the crisis we're living in," he said into my microphone. "There is no infrastructure here, man. You can see what we're doing. We are creating our own jobs just to get by."

"Until you can find something else," I suggested.

"You gotta get me outta here," Robinson said.

"I'm the wrong guy," I said. "I just got here myself."

"How long you in town?"

"Dunno. Couple of days. A week, maybe."

"Okay, man. Let's catch up later. Anything you need. You come find me, man."

"Thank you, Robinson. I'm just going to walk around a bit then."

"You just ask for Robinson!" he called as I moved off toward the wide plaza in front of the National Palace. "I ain't goin' nowhere!"

The next day Karla and I visited the CEP—the Conseil Electoral Provisoire—at its downtown headquarters, with its iron front gate still hanging askew after the recent grenade attack. A man hanging around out front introduced himself as Edouard François La Dolce. He was a bulky man and his right forearm was bandaged in white gauze. He said he had been at the CEP when the grenade went off.

"There were two pieces of metal sticking from my arm," he said proudly, "and some people had to help me to pull them out."

Inside the frenetic, sweltering offices of the electoral council, officials complained that they were understaffed and scrambling to train workers and get materials together. But they were confident they would be ready for the parliamentary vote on May 24. The problem, one official said, was that there was no institutional memory at the CEP. Each election cycle Haitians had to reinvent the entire wheel, from ballots to boxes to tables to set them on. No one ever stored materials for further use and no one seemed to work for the CEP long enough to keep track of any materials or systems.

Down the street, at the walled compound that served as headquarters for an opposition group calling itself L'Espace, dozens of campaign volunteers were busily getting posters and flyers together in the shade of the courtyard. Theirs would be a losing attempt at stopping Fanmi Lavalas from gaining a majority in the Senate, and I could see why they were having problems; for starters, they were working in the courtyard because the house that had served as their headquarters had been burned down. A mob chanting pro-Aristide slogans had in recent days stormed the place, chasing opponents over the back wall and burning anything they could. The blackened chassis of a car sat in the carport next to the large, gutted brick house.

"We blame Aristide for this," one well-dressed Haitian intellectual told me. "He was nearby when we were attacked and did nothing to stop this."

I was quickly getting a sense of how polarized Haitian politics were. At a pro-Aristide rally at a school that same day, Lavalas VIP and future prime minister Yvon Neptune denied Lavalas was behind the arson attack or the string of murders of Aristide opponents in recent weeks.

"The Haitian people know that the violence is not coming from them," he said. "It is coming from a tiny group of people that may be connected with the former repressive regime, and for political reasons they would not like the people to make their choice."

Aristide had become Haiti's first democratically elected president in 1990, ending nearly two hundred years of dictatorships, foreign interventions, and military juntas. But his tenure was short-lived: before the following year was out a cadre of army officers rose up and drove him into exile. Three years of brutal oppression followed. The new military junta, which had uncomfortably close ties to the CIA, terrorized and killed thousands of Aristide supporters and grassroots organizers. Thousands more fled into exile. President George H. W. Bush largely ignored the violence in his backyard. His successor, Bill Clinton, inherited tens of thousands of Haitians fleeing for Florida by boat as the Caribbean crisis worsened. Clinton eventually restored Aristide with the backing of twenty thousand U.S. Marines. The Marines stayed for three years, stabilizing the country and building and training a new police force that quickly became corrupted and politicized after they left.

As I stood outside the schoolyard a comfortable distance from the packed throng of Aristide cheerers, clumsily eating the first and most delicious Haitian mango of my life, I thought about the highly charged atmosphere in which I'd landed. Among Haiti's political classes there seemed to be no room for compromise or negotiation.

On the day of the vote I accompanied a young American election observer from a Washington-based, pro-Aristide nongovernmental organization called the Quixote Center. As the sun rose Haitians were already waiting patiently in long lines to cast their ballots. By midmorning they were baking in the capital's treeless streets. The American, who knew Haitian history well and was disdainful of U.S. policy in the country, steadfastly pushed me to agree that turnout was extremely high and overwhelmingly in favor of "Ti-tide" as Aristide was affectionately called by his supporters.

"It's amazing," she said.

"What's amazing?"

"This popular support for Aristide."

"We've only been to three polling stations so far."

"And what a turnout."

"Seems sort of hard to tell," I said.

The American eyed me warily.

"You don't work for the CIA do you?" she asked.

Polls closed late that day with no major incidents of violence or fraud, and then the wait began. When I returned from my mountain mishap with Ron Bluntschli, election officials announced that Lavalas had won eighteen of the nineteen Senate seats and thousands of local offices across the country. Opposition groups immediately cried foul, and days later so did the OAS electoral mission.

It turned out that the CEP had used a questionable formula to count the ballots. Its method gave the Fanmi Lavalas candidates an inflated percentage of the vote, allowing them to avoid runoff elections.

The OAS quickly released a statement condemning the procedure and calling for a recount. The CEP refused. Astonishingly, Fanmi Lavalas refused as well. Nobody was sure what was going on. Aristide, holed up in his multimillion-dollar mansion in the Port-au-Prince neighborhood of Tabarre, remained silent. The next several days were filled with backroom negotiations between the key Haitian players and the international community. But Haitians on both sides just dug in their heels. This was going to drag on for a long time. Glickman called and ordered me back to Mexico City.

On the night before I left Haiti I led Gaston, a young man from room service, up to the roof of the hotel for the last time.

"I promise this is it," I said.

"No problem," he said, fishing in the pockets of his over-sized black trousers for his keys. He turned the deadbolt and he unlocked the two padlocks.

"You should have just given me a key."

"There are three, sir."

"I'll let you know when I'm done," I said, and climbed alone to the flat roof. The sun had just gone down but on this evening the palace lights had not come on. Port-au-Prince was suffering yet another brownout. All of downtown was darkened as if bracing to be bombed. I set down my hard-case, flipped open the latches, and removed the phone. Then I lifted out the gray, three-paneled satellite dish and angled it in the direction I knew now by memory. I moved behind it and plugged in the cables, then switched on the phone and watched the green LCD screen as it searched for our bird in space.

I was feeling elated because I'd made it through my trial by fire on the foreign desk, filing stories day and night from this very roof with its clear shot at our satellite. And the reaction in Washington had been positive. But something was eating at me. Haiti had just celebrated free and fair elections that suddenly didn't seem so free and fair. Street celebrations had given way to crowds of worried faces. On top of that, being new to Haiti and not speaking Kreyol, my capacity to understand what was really going on had been severely limited. It made me think of Robinson the deportee again, struggling any way he could to get a handle on a culture completely alien to him. The big difference between us, of course, was that I was leaving.

The phone beeped. It had found the satellite and locked in on the signal. I picked up the cradle, dialed NPR's intake department, and bounced that evening's audio to D.C. My last story of this trip.

The next morning I left my hotel dragging my bags behind me when Robinson came running up. He was wearing the same clothes.

"Hey, man, let me help you."

"Thanks, but I've already called a cab."

"I don't want any money, man. I'm trying to get the hell outta here."

"I know," I said. "I'm sorry about your situation."

"And I know you ain't gonna help me," he said. "But telling another American at least makes me feel better, man. Even if you don't tell nobody yourself."

"I would like to tell someone," I said.

"Where do you live anyway?"

"In a hotel in Mexico City."

"Mexico? Man, I hear that's paradise."

"I couldn't really tell you."

"Well, I can tell you about this place," Robinson said. "Put this in your story, man. There is trouble coming. When I first got here I thought things couldn't be worse. I thought this place was hell. But now I know it ain't got there yet."

I sat down on my sat-phone case. "Is it that bad?"

"Man, if these elections come to nothin' and Aristide can't be president again, we'll get to hell quick. You ever live in a place where there's no hope? I mean no hope at all?"

My taxi pulled up.

"And then suddenly there's this tiny bit of hope that shows up?" he said, helping me put my bags in the trunk of the taxi. "And then, they come and take it away again?"

"Aristide," I said. I handed him five bucks.

"No one else."

Robinson slipped the five-note into his pocket and glanced around nervously like some Times Square ticket scalper. "If Aristide and his people don't get to rule this place," he said, "people will start wishin' he never showed up at all." He slapped the roof of the taxi as I climbed in. "Put that in your microphone, man."

At the airport I wound my way through check-in and into the crowded boarding area. In just a few hours I'd be back at my Mexico City hotel. I was looking forward to a long bath. I passed through passport control and sat on a plastic chair in the hot, crowded waiting lounge. While we were waiting I considered what Robinson had told me. Clearly he'd been projecting his own sense of hopelessness onto the Haitian people. Still, the mood on the ground as I was leaving was solemn and uncertain.

I slumped back in my chair, my eyes heavy with fatigue. Then suddenly my hands and feet went cold. Then numb. It was as if all the blood in my body were rushing to my torso. A tremendous surge of adrenaline caused me to sit up in my chair and gasp. I guzzled my remaining bottle of water, took some deep breaths, but felt no better. I began to feel faint. In a panic, I searched the crowd for a familiar face and I found exactly one: a blond American woman, a reporter for a newswire who'd parachuted in from Florida for the elections. I couldn't remember her name but I walked up to her.

"Do you remember me?"

"Yes," she smiled. "Hello."

"I feel very ill," I said. "May I sit with you?"

"Okay," she said, her expression uncertain. She had on khaki hiking shorts and one of those safari hats that won't lose its shape if you sit on it by accident. A matching duffle bag rested on the seat next to her. She moved it to the floor. It was a pointless gesture however, because soon I was on the floor too, writhing, unable to catch my breath. I clutched at the woman's bare calf with one clammy hand.

"Jesus," she whispered. "Get a hold of yourself."

"I think I need to go to the hospital," I said.

The woman leaned down very close to me and said harshly, "You are not going to a hospital in Haiti. Do you hear me? Do you want to get *really* sick? You are going to walk across that tarmac and get on the plane."

When the plane was ready the reporter helped me stand and walked with me across the baking hot runway. She smelled soapy and from elsewhere. I barely managed to get up the steps and into the cabin of the jet. I flopped down in my seat, sweating, short of breath, cold. Embarrassingly, I started writhing again.

"Gee, you don't look very good, Gerry."

I glanced up to see a white woman in a business suit staring down at me. It was Michelle Karshan, the international press liaison for none other than former president Aristide himself, standing in the aisle next to me, looking bemused.

"What are you doing here?" I whispered.

"Going home to Brooklyn for a few days," she said.

"Can't take the heat?"

Karshan was an American from New York who'd had a daughter with a Haitian man and lived most of the time in Port-au-Prince. I had no idea how she'd ended up working for Aristide, but I had been warned to hold my tongue around her, especially on the veranda of the Hotel Oloffson, a spectacular nineteenth-century Gothic gingerbread mansion and the favorite haunt of the foreign press. Karshan was, after all, the eyes and ears of Haiti's National Palace.

"You got any medicine?" I croaked.

"What's wrong?"

"I just need some painkillers," I said. Information on a need-to-know basis only, I thought.

Michelle dug up some kind of pills from her big black handbag and I took two and fell asleep. When we landed in Miami the wire reporter hurried off without saying good-bye. Michelle was nowhere to be seen. I wanted to thank her but it was no matter. I would have my chance when I returned in November for the presidential elections. To my dismay, as soon as I'd collected my bags my hands and feet started going cold again. I ran out to find a taxi as fast as I could. I had no intention of collapsing in two airports in the same day.

"*B*uenos días, *Señor Hadden*. How was the Caribbean?"
"*Una locura*," I mumbled to the concierge as I entered my hotel back in Mexico City. *Madness.*

I dropped my bags in my room and went straight across the street to a pharmacy on Durango Street. I described my symptoms to the pharmacist, showing him my sweaty palms. He listened impassively then dropped a book on the counter. It was a large encyclopedia of illnesses and their remedies. I began searching through the symptoms index and narrowed things down to about six hundred possible diseases. "Don't show this book to hypochondriacs," I said. Taking a gamble I bought a week's worth of some antibiotic, which didn't help.

Before long I found the house that would become the new Mexico City NPR bureau. It was a three-story row house with a two-tiered roof deck and billiard-table-green, wall-to-wall carpeting on the main floor. The walls themselves were painted two complementary shades of dark, mottled copper. The place was cast in eternal twilight. In one corner of the living room, beneath a long gilded wall mirror, stood a dusty black piano with a candelabra on it. Under the candelabra was spread an intricate white silk doily with gray lint collected around its frilly edges.

The place couldn't have been spookier.

"Isn't it lovely?" said Christy, the owner, as she showed me around. Christy sold car insurance and wore lots of Mexican silver. When she spoke, no matter what she was saying, she sounded like she was confiding in you.

There were white doilies under almost every object in the vast living room. It was an agoraphobic grandmother's paradise, a vault of delicate china and chintz couches. But at a thousand dollars a month it was a steal.

"I inherited this place from my mother," Christy told me. "You'd be the first person to live here since she passed away a year ago. It's in perfect condition."

In fact it was a time capsule. After her mother died Christy had simply sealed the house up. She hadn't even cleaned out her mom's belongings from the bathroom drawers and kitchen cupboards. Walking the house alone after Christy left I found the matriarch's personal effects still set neatly in their places: bottles of hairspray beneath the bathroom sink, a pink bar of soap in the bathtub soap dish, shawls and dresses in an upstairs bedroom closet, handwritten letters bundled in a desk drawer, a library card from the 1940s, a Mexican driver's license from 1968 bearing a black-and-white photo of a young woman with a pompadour hairdo. I decided I would pack all of these items away the next morning before unpacking my own things. I didn't want to feel as if I were sharing the house with someone else.

That first night I slept in the grandmother's bed, a single, listing box spring with a red velvet bedspread and an upholstered clamshell headboard wrapped in plastic. I did not sleep soundly. The next morning I found among my moving boxes one labeled BUDDHIST STUFF and carried it into a small annex off the guest bedroom. My idea was to use the space as a makeshift shrine room. I was looking forward to taking some time, even just a few minutes, to let the whirlwind of the last week play itself out in the solitude of meditation. A small moment without analyses or deadlines. This was how I imagined I might continue my practice in the midst of what I could now see was going to be my highly irregular lifestyle.

When my shrine room was more or less set up I padded downstairs to see if the matriarch had by chance left any coffee behind. As I rounded the landing that led to the living room I found myself looking down at a young man sitting on one of the old chintz couches. He was talking on the telephone. My telephone. He looked up at me without expression. I turned around and returned to my room and called Christy on my cell phone.

"Christy, *buenos días*," I said. "Listen, there's a guy on the couch using the land line."

"What? What does he look like?"

"Dark hair, medium dark complexion, maybe eighteen years old."

"Ah, that is Alejandro!" she said.

"Ah."

"Yes, that is Alejandro."

"And who is Alejandro?" I asked.

"He is the son of Concha."

"Who is Concha?"

"*Concha es la muchacha*," she said.

"I didn't know I had a maid," I said. "Or that her son comes to work with her and uses the house phone."

"I will have them install a second line," Christy said.

"Well, I'd prefer that he not use any line," I said. "Surely he can make calls from his own house."

"But that is his house," Christy said. "Concha and Alejandro and Diego live with you, in the upstairs chamber off the office." I remembered seeing three small beds in a tiny room on the roof, but I hadn't thought much of it, given that the rest of the house had an air of still being inhabited. I had figured it was just another room yet to be cleaned out.

"Here?" I said. "You didn't tell me a family was living here."

"But I asked you yesterday if you wanted a maid," Christy replied.

"Right. And I said I needed to think about it."

"Alejandro and Diego were born in that house. They have been living there twenty-five years."

"Christy," I said, "I can't live with a family. I work weird hours. I need the phone lines. This is also an office. I need my own place. I wish you'd told me."

There was a long pause. "Don't worry, then," she said finally. "I will tell them to leave. I agree with you. Since my mother died last year Concha and her boys have stayed on. But there's been no need, no work for them. It is time for them to go."

"Hold on," I said. "I don't want to throw them out on the street either."

"Oh no, Concha's family lives just outside Mexico City. They can go there. They will not bother you."

"Jeez, are you sure?"

"Yes. Their time in the house is now finished."

I hung up feeling embarrassed to go downstairs. And it wasn't just because I was still in my underwear. Concha and the kids didn't know it yet, but my arrival was going to spell the end of a lifetime in that house and I felt badly. I got dressed and hurried guiltily down the back steps and out the door. I bought the morning paper and found a sidewalk café at a leafy intersection of trendy restaurants and began skimming the local news for clues to stories of interest, stories that might work for radio. I wandered back to the house early in the afternoon and found Concha, Diego, and Alejandro all upstairs in their small room, packing clothing. So they'd already heard from Christy.

"*Hola,*" I said.

"*Hola.*"

"*Hola.*"

"*Hola.*"

"*Me llamo Gerry,*" I said.

No one responded.

"Listen, I wouldn't have signed the lease if I'd known you were here," I said.

"*No se preocupe, señor,*" Concha said. Concha was short and round and dwarfed by her two tall sons. She wore a blue maid's smock over her dress. "Don't worry. It is time for us to go."

"You don't have to go now, this second," I said.

The two sons said nothing as they continued packing, but they'd heard me. I know this because they did not leave that night. Nor that week. After a month I began to suspect that they'd simply unpacked their bags again. I began to get angry.

Alejandro, Concha's younger son, was eighteen.

"I run a one-man business selling telephone calling card services," he told me.

"I see."

One morning a few weeks after I'd moved into the house I went to his office. By accident. I'd seen him go through a little door into the empty space under the main staircase leading from the garage to the first floor. I hadn't noticed it before. I followed after him.

"*Hola.*"

"*Buenos días.*"

"*Con permiso,*" I said.

"*Pasele.*"

I stepped inside. The space under the stairs was cramped and dark, lit by one bare light bulb and the purple glow of a fish tank that gurgled in one corner. There was just one fish in it.

"That's a Siamese fighting fish," he told me. "*Es muy chido.*" *It's very cool.*

"They live alone?" I asked.

"They cannot live with others," Alejandro said. "They kill the others."

"Well, sometimes we need our space," I said. "What is this, your clubhouse?"

"This is my office."

"This is where you work? You run your business from under the stairs?"

"As you say, *señor*, a man needs his own space."

"*Orale,*" I said.

"*Pero no se preocupe, Señor Gerry,*" he said. "I am no longer using your phone. I now conduct my business from the pay phone on the corner."

"What about your cell phone?"

"Unfortunately, there is no coverage here under the stairs."

"Well," I said, at a loss. "I won't keep you then. I've got to go to my office as well."

I climbed the stairs to the roof and found Concha once again ironing shirts on top of the desk where my new assistant, a recent university graduate named Fernando, worked. She'd set his various stacks of papers on the floor.

"*Buenos días.*"

"*Buenos días.*"

"*Traigame sus camisas, señor,*" Concha said.

"Oh no, you don't have to iron my shirts," I said. "Fernando should be here *any minute.*" I didn't want Concha or her sons doing anything for me. It would just give them pretext to invade my space further. Or to stay longer. The situation was uncomfortable enough. At lunchtime each day Concha would cook for her two sons, taking over the kitchen from noon until two. Then they'd all retire for a two-hour siesta in their bedroom off my office. If I was home I was forced to work more or less in silence from two until four for fear of disturbing their slumber.

That day I called Christy again.

"They're still here."

"That cannot be," Christy said.

"Can you hear that?" I said, holding up the phone.

"No."

"Two of them are snoring. Which is not a problem. I snore. But not in somebody else's office."

That evening Christy showed up at the house. I was downstairs watching the news on television. Christy and Concha and sons all marched upstairs. Half an hour later they marched back down. My roommates were all carrying suitcases. They followed Christy out the front door without so much as directing a word at me.

The next day I was out most of the afternoon trying to work out my journalist's visa with the Interior Ministry. When I got home I looked inside Alejandro's office. Everything was gone, including the fish tank. My neighbors on Jojutla Street eyed me with open disdain after that. I knew what they were thinking: *Gringo, que te chinges. Go screw yourself, gringo. There goes the neighborhood.*

With Concha and company out I started in on what was going to be the biggest story of my tenure. It was the story of the man who was likely to become Mexico's next president and dramatically improve relations with Washington, boosting trust and cooperation on many fronts, including the explosive issues of drug smuggling and illegal immigration. His name was Vicente Fox Quesada, and on the eve of Mexico's elections his press team invited me aboard a private jet the businessman-turned-politico had borrowed for his final few campaign stops.

We met at the airport. Accompanying Fox that morning were his prim campaign manager and suspected lover Marta Sahagún, a handful of bullish men I assumed were bodyguards, and Fox's teenaged daughter, Christina, chewing gum in a pretty white suit.

The plane was hot and cramped and the mustachioed Fox, standing six-foot-four, practically had to crouch in his seat while he answered my questions about his plans for governing were he to win the vote on July 2. He was wearing a dark suit and the customary black cowboy boots that had become a symbol of the down-home rancher image his handlers were trying so hard to cultivate.

"My style is totally of the people," Fox told me. "I am not a career politician. I'm a businessman, a happy family man, a farmer."

Fox had in fact already cut his political teeth, serving one term as governor in the central state of Guanajuato. Huge sweat marks stained his armpits and mine, and I suppose those of the big guys sitting silently and at some distance, though they did not take their suit coats off. Only Marta seemed impervious to the sweltering conditions on board, her hair and makeup holding up rather well. By way of introduction she explained to "Chente" that NPR was an influential and respected news outlet in the States and that many Mexican Americans listened to public radio. I wasn't sure if she said this to flatter me or to nudge Fox into interview mode. He nodded thoughtfully. Like many Latin Americans, Fox, it was clear, did not know National Public Radio.

Today would be my one chance for an up-close glimpse of the man on whom millions of Mexicans were pinning their hopes to end more than seven decades of rule by the Institutional Revolutionary Party, or PRI. The PRI had been famously labeled a "perfect dictatorship" by Peruvian writer Mario Vargas Llosa years earlier because it always managed to look like a democracy while functioning more like a giant, menacing mafia. Buying off opponents. Blackmailing enemies. Otherwise silencing challengers. For the minority of Mexicans running the PRI machine, life was easier than it might have been under a democracy with transparent institutions. Until Fox came on the scene and began scoring big points in the polls, Mexicans had seemed resigned to living ad infinitum under the weight of institutional corruption and local bullying. Now, glimpsing a real chance for change, they were suddenly clamoring for it.

Fox was an amicable man, earnest, unpretentious, and physically awkward in the way large men in business suits tend to be. He had his speaking points down and delivered them calmly and without excessive sparkle, and he impressed me greatly. With his daughter sitting nearby and listening in, I felt at times more like a guest at a family gathering than a reporter on the clock.

But this was no picnic. Fox had less than a month to solidify his lead against PRI candidate Francisco Labastida, a drab party lifer who knew little about charming a hungry public but who had the PRI's far-reaching influence to guarantee him votes en masse.

"Mr. Fox," I asked, "how, as president, do you hope to rid Mexico's vast public sector of corruption?"

"It will not be easy," he said. "But we'll start at the top, ensuring that top civil servants set a good example. But it will take time."

"What will you do to reduce the number of Mexicans who cross the U.S. border illegally in search of work? Is that even a good idea, from your perspective?"

"I want to revisit the terms of the North American Free

Trade Agreement," Fox said. "Since NAFTA, salaries in Mexico have not risen in real terms. As long as Mexicans keep earning six dollars a day at home, and sixty dollars a day in the States, we'll never end illegal immigration.

"I have a vision of North America functioning like the European Union, with borders open to the movement of people as well as goods," he said.

We landed on a blazingly hot runway in the ugly oil town of Poza Rica, along the Gulf Coast, where a crowd of several thousand had amassed before an outdoor stage. They were waiting anxiously to hear Fox's promises of a Mexican social and economic transformation. We arrived at the open field in a small bus and piled off into the bright sunlight. Fox, broad-shouldered and a full head taller than me, waded quickly into the crowd. About a hundred yards separated us from the grandstand. I struggled to keep up, both hands held aloft, one grasping my shotgun microphone, the other a minidisc recorder, rolling digital "tape," not wanting to miss any chance for good sound. At any moment a hopeful supporter might shout out some impassioned plea for help that would bring my story to life.

Please, Don Vicente, a chicken, just one chicken.

Por favor, Mr. Fox, I need five thousand pesos for my tortilla stand. Do not forget me.

It also crossed my mind that an assassin would need just seconds to place a gun to Fox's head and change history. It had happened to presidential candidates in Mexico before.

As we crossed the distance to the stage the crowd began to surge, pushing and pulling, teens and mothers and sweating *mestizo* oil workers reaching out to touch Fox as he strode past. A sea of dark eyes under a sea of stained palm hats. I nearly fell twice but managed to keep my deck rolling. By the time we reached the grandstand I was drenched and greatly relieved to be out of the multitude.

Fox rolled up his sleeves and began his speech, considerably more animated now than during our interview. The crowd went wild. Fox rattled off a series of platitudes revolving around his principal slogan, *Cambio.* Change. He threw in some street language to bolster his credentials in that decidedly working-class crowd.

"The PRI has been fucking you over all of your lives," he bellowed. The crowd ate it up. "Tell them what you think of that on July 2!"

It was over in twenty minutes. Then we thundered across the state in a caravan of SUVs to the ruins of Tajín, north of the port of Veracruz. Tajín is a vast pre-Columbian complex of stone temple foundations, houses, and open courtyards dating back to the first century. We passed a playing field where contestants long ago competed in a sport in which you tried to knee a small ball into a hole in a stone wall. One unsubstantiated legend has it that priest-referees would execute the winners as a reward for their superior sportsmanship. More clever players who tried to lose on purpose were also executed.

Our guide was a young Mexican man with a thin beard and a New Age rain stick that made a pleasing rattle when inverted. He made us lie down in the shady grass to one side of the temple ruins. "If you close your eyes," he intoned, twirling his stick, "you will hear the footsteps of our ancient ancestors as they journey through their netherworld." I wondered if the guide had to do this for everyone or whether this was a special presentation for the guest of honor. As our mystical journey into the past began I fell asleep and caught almost none of the adventure. Then Marta was nudging me.

"Sit up," the guide was saying, "and open your eyes. Now, who heard the footsteps?" A few hands went up gingerly; slowly, "Chente" raised his.

"And how do you feel now, Señor Fox?" beamed the pleased-looking guide.

"Calm," said Fox, smiling like a self-conscious student in a classroom.

Though I can't be sure, I suppose that that was one of the last times Mexico's future president would think to describe his life in those terms. On July 2, 2000, Vicente Fox won the presidency, making modern history and kicking off a five-year term in office that would consist largely in banging his head against a wall.

I thought I'd gained my privacy after Concha and her sons moved out, but I was mistaken. One Friday night, not long after they'd left, I invited a group of Mexicans over to the house after a late party. My new life was getting into full swing. We sat on the upper roof terrace looking out over the lights of the Condesa, drinking the woody, aromatic mescal I'd found in an old unlabeled bottle in the kitchen. In the near distance, the Castle of Chapultepec was lit up impressively by hidden lights. The eighteenth-century fortress and surrounding park occupied a wooded hill and had served as a presidential palace, a military academy, and, now, a history museum. In 1847, during the Mexican-American War, it was a training center for cadets. That was the year the U.S. Marines arrived. When they laid siege to the castle the Mexican army retreated except for six student soldiers. *Los niños heroes*, the child heroes, defied their superiors and stayed behind to defend their school. They were all killed.

Legend has it that the final surviving cadet leapt from one of Chapultapec's high windows wrapped in the Mexican flag to save it from falling into enemy hands.

Now here I was, I thought, a gringo living in the shadow of that majestic symbol of Mexican defiance toward my country. How much had relations really changed, I wondered. It started to drizzle so we moved inside to one of two rooms on the roof, a spacious library packed with books ranging from a collection of Star Trek paperbacks to Hitler's *Mein Kampf* to Bertrand Russell's *The Problems of Philosophy*.

I had converted the library into my office, installing high-speed Internet and a local area network that gave me a real-time studio-quality audio link to Washington. Now, when I spoke to HQ via the LAN line from three thousand miles away, it sounded like I was sitting right in one of NPR's Massachusetts Avenue studios.

Early each morning, I would touch base with my editor, Paul Glickman. Over our respective cups of coffee we'd hash out what stories we wanted to cover, how yesterday's story had sounded, and so on. At any point in the day we'd also edit my upcoming pieces.

This consisted of me turning in a written script of a story, including a suggested lead or introduction for the host to read. The scripts looked much like movie scripts, with each "character" identified along with what he or she said.

Woven throughout were indications of what background sound, or ambience, ought to be supporting the scene—painting the scene—in the minds and imaginations of listeners. Without it—a thousand geese honking in a French feeding house, the quiet plink of water dripping from a Bedouin water pump—your story would be nothing but a string of talking heads. That is boring, and will cause listeners to drift away or switch you off. In public radio feature reporting, a string of talking heads equals death.

My second desk no longer had Concha's sons' shirts piled on it, which made life easier for my part-time assistant, Fernando. Fernando was short and slight and wore a thick brown ponytail. He came recommended to me via a Mexican friend back in Seattle. He was bright and educated, having majored in English literature at Mexico's fine National Autonomous University. He was also chronically late, which maddened me to no end. Next to Fernando's desk was a doorway leading to a second room—the now vacant chambers of Concha and her sons. During the course of my late-night mescal gathering one young Mexican man in thick glasses kept glancing toward that door.

"Do you see that too?" he asked me.

"No."

"But you definitely have ghosts," he said, in a way that implied that I already knew what he was talking about. I wondered how this guy could spot ghosts when I was sure he could barely read a stop sign through those lenses.

"Yes," concurred a young Mexican woman in ripped jeans and a big black sweater. "It is clear."

"Really, how can you tell?" I asked. Now I was getting intrigued. Somehow the idea didn't bother me so much. I'd always wanted to see a ghost, just as I'd always hoped to catch a glimpse of a UFO.

"You can tell when your house is haunted," said the guy.

The comment served as a cue for my guests to dive eagerly into their own experiences with the occult. It seemed that everyone had a tale to tell. And everyone agreed that I would soon have some. Some ghosts, I was told, were nice.

"Sometimes when I'm sad and I take a bath, my grandmother shows up," one woman said.

"She shows up in the water with you?" I asked. "I don't think that would make me less sad."

"She's not your grandmother."

Toward dawn my invitees straggled out, still telling tales of the dead. By that point I was drunk and really spooked; as I shut the front door the hairs on my arms stood up. The house suddenly felt bigger and darker. I decided I had better deal with this straightaway. Then I stumbled upon what I thought was a brilliant idea. I could try to adopt a Buddhist position on ghosts. It would be yet another way to continue practicing. Buddhists view ghosts as beings experiencing some kind of confusion. Otherwise, goes the reasoning, they wouldn't be clumping around in your house or yodeling under your bed. They might scare you, but you're supposed to show them compassion. If they go bump in the night, you invite them to bump some more. You ask them what they want. You tell them they can stay as long as they want, that there's room for everyone. And here's the tricky part: you're supposed to mean it.

I knew in my heart that I didn't mean it. If there were ghosts I just wanted them to go away. But I had to give it a try, as phony as I felt. First, I decided to burn up some of my fear by climbing to my office via the narrow, unlit back stairwell, the old servants' route from the roof that led directly to the kitchen and garage, bypassing the third floor with its two bedrooms and bath. I nearly had to hunch over as I felt my way up the chilly staircase. I used every ounce of willpower not to bolt. I passed the kitchen, in shadow and quiet except for the hum of the refrigerator. Two more floors and I stumbled over to my desk, utterly unnerved by the ascent. I sat down and raised my glass to no one in particular and then let loose a slurry toast to the spirits that might have been listening.

"I dunno what you've been told," I said, "but here's the deal. You can absolutely stay." I paused for dramatic effect. "Every last one of you. I got four stories of house here. There's plenty of room. So take a load off." To show them what I meant I swung my feet up on the desk and leaned back in my chair. I felt better and sufficiently silly then. "Peace on earth," I said. "And goodwill to all ghosts and ghouls and goblins."

I went downstairs and climbed into my new king-sized bed—I'd retired the clamshell—and pulled the white mosquito netting closed around it and fell asleep. Not long thereafter one of my closet doors flew open. It was the one that I kept locked. When I rolled over and sat up it was still swinging slightly on its hinges. I was nearly weightless with fright, and exhilarated. The room was cast in the pale, yellow glow of a streetlight. I took a deep breath and started clapping.

"Bravo," I said, "You did it. You scared me." My words sounded hollow, like the cheap reverse psychology I wanted to avoid. But this idea of welcoming what scares you was alluring. I was terrified but I wanted to get closer to whatever might have been lurking around. "Wait till I fall asleep," I said, "and please do it again."

I lay back down and tried to sleep. But it was impossible, with the closet door standing open like that. I kept imagining someone's grandmother leaping out to comfort me.

I was reading in a local paper that a group of Mexico City intellectuals was to hold a symposium on the British philosopher Bertrand Russell.

"You like Bertrand Russell?" I asked Fernando.

"He's great," he said nervously, bounding up the steps to his desk.

"For the love of Christ, could you try to get here by nine a.m.?"

"Sí, sí. Perdon."

Fernando's main job was to help me set up interviews, especially with people from the Mexican government. You needed a lot of persistence and a firm handle on all the nuances of Mexican Spanish in order to win over the suspicious bureaucrats unused to explaining anything to anyone, much less to a foreign journalist—and an American one, to boot.

"Did you know that your desk isn't really a desk?" I asked him, to see if he was familiar with any of Bertrand Russell's ideas.

He said yes, effectively ending the conversation. We spoke no more of Russell that day.

At lunchtime Fernando went home. I worked for a few more hours then turned off the lights and went downstairs.

The next morning when I climbed back up to the office I found a copy of Bertrand Russell's *The Problems of Philosophy* lying on my laptop computer. I picked it up. It was a slim volume, in Spanish, with many notes in the margins and sections underlined in black pen. I assumed Fernando had brought the book over and left it for me, based on our brief conversation the day before. I opened it and began to search for any particular section he might have wanted me to read. A folded page, a bookmark, something.

I was reading it when Fernando came bounding up the steps again, six minutes late.

"Fernando, I know we're in Mexico but my boss is in D.C. You've got to get here on time. My phone starts ringing at nine a.m."

"Sorry," he said, smiling guiltily.

"What are you reading?" he asked.

"What part did you want me to read?" I asked.

"What part of what?"

I tossed him the book. "Of this."

"This isn't mine."

"You didn't leave this for me?" I asked. "I found it this morning sitting on my computer."

"No," Fernando said, "I left before you yesterday, and I didn't come back in the evening."

In fact no one had come over that evening. I'd been alone in the house.

"So where did the book come from?"

"Maybe from the shelves," Fernando said, gesturing to the wall of books behind me. The hair on my arms stood up. We began searching the shelves for clues. About ten feet from my desk I located a handful of dust-covered Russell volumes clustered together. One was missing. The space it had occupied was free of dust, indicating that a book had recently been removed. The book I'd found on my computer fit perfectly into the spot.

"You must have taken it down to look at it after we spoke about him," Fernando said.

"No," I said, "I didn't."

"And was anyone else here?" he asked.

"No," I said, beginning to smile.

"Why are you smiling?"

"Because of the ghosts," I said. "They don't just slam doors. They deconstruct them! And they're generous!"

"What?"

"They heard us express interest in Russell, so they left me the book. They could have left me, I don't know, a bag of rotten *chayotes*. A bloody iguana's head. But they gave me a book."

Fernando nodded pensively

"Your thoughts?" I asked.

"I'm realizing that I'm not chronically late after all."

"You are indeed chronically late, *colega*. You have Chronically Late Syndrome."

"But if our desks are not really desks," he said, "then showing up late to them becomes impossible."

"Give me that book back," I said, laughing. I set it in its original space on the shelf. "Maybe this wasn't such a nice present after all. It's filled with too many dangerous ideas."

"*¡O*rale, *cabrón!*" I yelled over the loud mariachi music. "*¡Estoy listo! ¡Dale!*" *Go for it, you son of a bitch! I'm ready. Bring it on!*

Tomas the Shocker turned the dial.

At first it felt like tiny insects were squirming in the palms of my hands. Then the muscles in my hands began to contract involuntarily around the two metal rods I was clutching. My knuckles turned white and my elbows began to buzz. Tomas the Shocker slowly turned up the voltage on the cluster of batteries wired to the rods. Fifty volts. Sixty, sixty-eight . . .

"*¡Sigale, sigale!*" he shouted as my face turned purple. *Keep going, keep going!* I wanted to, but the discomfort was turning into outright, vein-throbbing pain. As the current arced across my chest and my jaw clenched shut I thought, I have Vicente Fox—and only Vicente Fox—to thank for this.

President-elect Vicente Fox, that was.

On July 2, 2000, Fox won Mexico's national elections,

knocking the PRI from its worn and wizened seat of power in the fairest vote in the country's history. I was thrilled. First, I liked Fox and believed he was just the sort of straight shooter who might get a handle on Mexico's big challenges, most of which seemed to stem from the systemic corruption that pervaded all levels of government.

Second, I now had free rein to go out and take part in all sorts of arcane traditions and bizarre rituals, and then do stories on them. For example, this ritual—or was it a pastime?—which pitted your body against what amounted to an industrial battery charger.

Following Fox's win my editors simply cut me loose. Mexico was now of the highest interest. "Tell us things about Mexico that we don't know," Jenkins told me. So I tried. I did stories on the country's bullring surgeons, specialists in treating horn wounds; on Mexico City's dreadful lack of seeing-eye dogs for the blind; on a thriving witches' market where you could buy love potions, and for enough money, said one warlock with a wink, elixirs of death; on Mexico's last "train yard village," where families lived inside abandoned boxcars alongside the tracks and sent their kids to an official boxcar school. And on and on. And now I was pretty sure that most Americans didn't know about this electric shock game—even though it had been around for half a century. It was completely unregulated, yet legal, and each shocker was free to build his own contraption as he pleased.

"I've been shocking people for thirteen years!" Tomas had told me as I was deciding whether I would take the ride.

Tomas was a stout man in his midthirties, and he wore his shocking apparatus like an accordion across his upper stomach. To attract clientele he wandered from bar to bar clacking the metal rods together.

"Let me get this straight," I asked him. "You electrocute me, then I give you money?"

"In all these years no one's ever been hurt!" Tomas shouted, "Why, it's even good for the circulation! I shock myself at least once a day! Usually in the morning! To wake myself up!"

We were in a crowded bar near the Plaza Garibaldi, the heart of downtown Mexico City's mariachi scene. Outside, trios, quartets, and larger ensembles, dressed in colorful, embroidered suits, crooned out ballads for tourists and lovers able to part with a few pesos. In the bar the ambience was festive. People were still riding the euphoria following Fox's victory.

"Business is great!" said Tomas. "People are animated! It's like they don't believe the PRI is gone—they ask me to shock them to make sure it's not all a dream!"

Across the border, the Americans were just as pleased as ordinary Mexicans. At last a Mexican president unabashedly supportive of the United States, a guy who had run Coca-Cola's marketing in Latin America, a leader who understood the American mentality and our myths—the self-made man, anyone-can-be-president—and so on. Future American president George W. Bush was one of the first leaders to call Fox to congratulate him. And as I'd expected,

I was on the radio day and night reporting on what Fox's victory might likely mean for U.S.-Mexican relations. The post-election party lasted for more than a month, and it presaged what was going to be intense American interest in all things Mexican over the next several years.

Even Jesse Helms came down to underscore the point. The senator had spent his entire political life disparaging Mexico as a corrupt and troublesome neighbor. Mexicans loathed him. I was ecstatic.

"Would you lend me your shocker?" I asked Tomas.

"*¡Híjole, señor!* I don't think that would be a very good idea! You have to be experienced!"

"I just mean the rods."

"*¡Orale!*"

"Like this?" I say, squeezing them. "I don't have to stand in water or anything?"

"No, my friend! And remember, this is not a macho contest. In fact, women usually do better than men."

I leaned toward the bar and tossed back my shot of tequila, then turned on my digital recorder. Long live Mexico, I thought. Long live Vicente Fox Quesada. I handed the microphone to Tomas and asked him to point it at me.

"*¡Orale, cabrón! ¡Estoy listo! ¡Dale! . . .*"

The man in the red headdress brought the machete careening down toward the freaked-out goat's head. Wonder of wonders, I thought. Just a couple of days earlier I'd been electrocuting myself to the music of mariachis. Now I was back in Haiti, ostensibly to cover the presidential elections. But once again the Bluntschlis had dragged me off the main story.

"Listen," Karla said, when she picked me up at the airport. "I know the vote's in a couple of days, but we've got to leave Port-au-Prince and I want you to come with us."

"When?"

"Tomorrow night."

"That leaves us two days to set up the elections, then."

"Actually, just one. This person lives a long drive away from Port-au-Prince. We'll need to leave after lunch tomorrow."

"Okay, so we have a day and a half. Let's get on it."

We skipped my hotel check-in—I'd booked a room at a posh hotel up the mountain called the Montana—and set out directly to gauge the scene on the streets. The atmosphere was identical to the mood in May. People were scared. Small bombs were exploding around the capital. Two people had been killed and several injured.

In one market a bare-chested vendor wearing broken glasses held together with twisted wire told me he'd nearly become a victim himself.

"Somebody hid the bomb in a handbag," he said. "It exploded right here at the market. The culprits hid it on a table covered with secondhand clothing. A young woman was picking through the garments when the device detonated. It tore her hand right off.

"They are trying to scare us away," he said, "so that the little people don't have a chance to participate and to vote for someone who can help us into the future."

The man was referring to the little people's champion, Jean-Bertrand Aristide. Aristide had been keeping a very low profile during the campaign, in part because he could. He was running uncontested.

Haiti's opposition parties were boycotting the presidential vote. And both the United States and the Organization of American States were refusing to send election observers. Everybody was still mad about the funny May legislative elections.

Consequently, as election day neared there were almost no signs of an election. No posters, no rowdy rallies. Just the juxtaposition of the occasional bomb blast with the silence from Tabarre, Aristide's multimillion-dollar mansion on the outskirts of the capital where he was lying low. Finally, that night, the candidate spoke publicly. He denounced the violence that his supporters were blaming on opposition thugs and that others, in turn, blamed on Aristide's followers and over which no arrests would ever be made.

"It is with much sadness in my heart that I bow before these innocent lives," Aristide said on Haitian radio in his characteristically slow and solemn cadence, "while we wipe the tears from the eyes of the victims' parents with the handkerchief of peace. The suffering of one of you is the suffering of all of us."

His prose, eerie and purple, nevertheless lifted many Haitians' spirits. Their man had spoken. On Saturday the streets were crowded with shoppers again and the pall of election dread seemed to have lifted. Roadblocks of old tires and piled junk that had begun appearing as a measure against violence now took on new purpose, serving to delineate makeshift soccer fields for neighborhood kids.

At midday we took Highway 1 from Port-au-Prince east toward the Dominican Republic. This was the highway that tens of thousands of Haitians took each year to work on the other side of the border. And the route they were forced to take home when Dominican politicians decided periodically to blame them for the DR's woes and expel them. It was a

straight, two-lane country road, unlit, lined with the detritus of villages, wandered at all hours by the ill-shoed and the
barefoot.

For this trip I was not going to need my recording gear.
In fact it was forbidden. So I left my minidisc recorder and
my microphones in my bag in the Bluntschlis' living room.
I checked to make sure everything was secure, including
my elongated "shotgun" microphone for capturing sound at
great distances. Because of its odd length, I'd built a custom carrying case for it out of a tube of cardboard cut to
size. I'd capped both ends with plastic then wrapped the
entire thing in several layers of silver duct tape. The tape
held the whole thing together and also kept the mic from
getting wet. When I was on the move I'd fasten the tube to
the bottom of my gear bag and not have to worry about it
getting in the way.

All of my gear was safe and sound.

I joined Karla and a young dreadlocked friend of hers
named Djaloki on the front porch. We climbed into a small
car and puttered out of the capital.

"We are going to a Voodoo ceremony for Mr. Jean-
Bertrand Aristide," Djaloki told me. "The people want him
to win."

"Seems like overkill," I said, "since he's running against
nobody."

"The people want him to win and *stay winning*," Djaloki
said. "To stay in power for the full five years. And not to
forget the people."

"I can understand the first concern, but is the second one really a worry?"

Djaloki changed the subject. "Tonight is also the one-year anniversary of the opening of this Voodoo temple. The priest who started it has quite a story to tell. One foot in two worlds, it would seem. He is from your hometown."

The way Djaloki told it, the priest was a Haitian man whose family had moved to New York when he was just two years old. He grew up there, went to school there, eventually became a rich man there. Then one night, asleep in his luxury Manhattan apartment, he had a dream. A Haitian deity came to him with an unsettling bit of news. The businessman's destiny was not to play out his days in the Big Apple but in the Haitian countryside, in a small village in the southeastern hills. He was to go there and found a Voodoo temple.

His people needed him.

"Check it," Djaloki said. "The next morning the guy wakes up feeling all disturbed. He tries to forget the dream, but it returns the next night. This time he pays attention. He buys a ticket to Port-au-Prince and hires a car and drives more or less to the point he can recall from the dream. He has no memory of Haiti and only a vague sense of his own Haitian identity. When he arrives at a certain village he stops. He walks around a bit. People start coming around to check out this well-dressed dude, including the village leaders. Through his interpreter he explains to them his dream. He tells them his name.

"'We are your cousins!' the Haitians exclaim. 'Your grandfather is from this village. Though he left as a young man many years ago.'

"So the man flies back to New York and sells what he owns as fast as he can. He begins studying Haitian Kreyol. He sells his apartment, his assets, his business. When he is free of his American entanglements he takes on these new ones in Haiti. He follows the deity's prediction and becomes a priest's novitiate, passing through all of the rites and studies. When he's ordained, he begins work on a Voodoo temple in the village of his grandfather. He finished it a year ago. You're about to see it."

We arrived at dusk and entered the house of worship. It was round and open on all sides save for a low cement wall. The ceiling was held up by brick posts and wood poles and against the only section of floor-to-ceiling wall sat the priest's chair. And that's where we found him, seated before his people.

"Hi," he said. "Welcome. Tonight we're having a party."

"Thank you," I said. I told him I was from New York too.

"Then you've come a long way to reach this place," he said. He motioned for me to sit in a chair at his side as the festivities began. Before us couples danced a kind of square dance around the central pole of the temple to music blaring from a radio. Kids ran around with plates of beans and pork.

"So this is for Aristide," I said to the priest.

"Yes," he said. "He needs all the help he can get."

"Why is that?"

"The Americans are against him."

"Is that so?"

"They do not want him to win. If he does he will pull the U.S. into tricky waters. As before, Washington will try to pressure Aristide to liberalize the economy. They will withhold aid and give audience to the opposition. But this might push the country toward political collapse."

"The U.S. could also ease up on Aristide," I said, "to try to work with him."

"The problem is that the Washington establishment finds him undesirable. I know this. Aristide does not hold dear their notions of free trade. He does not want to open Haiti to competition from U.S. companies. He knows that Haiti would be swallowed whole."

"How far can the Americans push him before he breaks?"

"There is no longer an army to oust Aristide," the priest said, "but this political standoff could run the country into the ground. But I tell you, the real problem is the uncertainty. It always is. Nobody knows who might fill the vacuum if Aristide were to be forced out again. And if you know Haiti's history, you don't want to imagine the scenarios."

As the night wore on the young, the drunken, and the women slowly drifted away, leaving about twenty men sitting in the temple under the harsh glare of bare light bulbs. It was 2:00 a.m. and I was starting to nod off myself. A few others straggled away. At some point the priest stood up and motioned for me to follow him. He led those who remained out back behind the temple toward a small wooden shack.

He opened the door and held it for us as we filed inside. It was dark in the hut and filled with the smoke of some fragrant wood. There were other men already inside and they were shifting back in forth on their feet in time to the percussive sounds a couple of young guys were making with their mouths. We were in there for a long time. The smoke was making me light-headed, or perhaps it was the music. We were swaying in unison back and forth like sea grass, and everyone was adding something to the vocal beat.

Then the door burst open. We went jogging out into the now cool air and around the side of the temple, and there in the clearing in front of us burned a huge wood fire. The women of the village circled the blaze. They were dressed in blue-and-white dresses with white head scarves. They were holding hands and singing. They opened their circle and we entered and formed a smaller ring within. The music was lovely and I wished I'd had my recorder with me.

Then the circle opened again and a man dressed in billowy red silk pants and a red shirt slid past us. His head was wrapped in red cloth. He held a machete loosely in one hand, in the other a length of rope tethered to a very nervous goat. The man danced about in a forceful manner, striking the machete against the hard earth, switching direction, writhing, leaping, all the while hauling his reluctant dance partner along. The singers held steady, sparks rose into the dark night and drifted over us like fireflies. The man with the machete turned the tether two or three times around his muscled wrist to secure it, then raised the machete into the air. Then he looked toward the sky, and I saw that it was the priest. I barely recognized him. His eyes were wide open. For a moment in that yellow, unreliable light he seemed like a man transformed—half here, half purchased among his deceased kin in some eternal diaspora.

Then he brought the machete down in a furious but wild arc. The blade, in his unsteady hand, missed its mark, instead striking the goat squarely on top of its skull, between its ears. The sound of it: an aluminum bat pinging a baseball into deep left field. Man and goat looked at each other. And the otherworldly priest came straight back to earth. In his eyes now was a mix of frustration and embarrassment.

The goat's gaze was all terror and awareness. It bleated once, loudly and for several seconds. The sound cut through the music more effectively than the blade meant to cut through the animal's neck. The man, no longer quite in control, waited out the animal's shrill retort like a chastened child. But the goat could not bleat forever, and when it finished the man leaned forward and brought the machete down again, unluckily striking the same spot. This time the goat made no sound but fell upon its forelegs dazed, probably out of this world already. The third blow found its intended mark. The goat's head fell away and the man grabbed it by one stubby horn. The blood sprayed from the torso in a mist made visible by the firelight and pooled in dollops on the earth, forming dark shadows. The man tried to dance off, but the head was still attached to the animal's steaming body. Now, like a regular butcher, his brow furrowed in concentration, he thwacked at the remaining tendons and skin holding his sacrifice together. In short notice the head was free. He jumped up, finally able to resume his dance, his head thrown back as the goat's head swung high and low. Around the circle he went.

The muted bell peal of machete off bone was still ringing in my ears. The goat's last bleat like the frightened complaint of any man who perceives death at hand and his own powerlessness to avert it. Then Djaloki was yanking me out of the circle by my arm, back into the dark. The air away from the fire was cooler and the sky was clear and moonless.

"Man," he whispered. "I'm going to start a new Voodoo movement." He shook his head. "No more of this animal sacrifice bullshit."

Before I could ask why, he added, "I'm like you. I'm a vegetarian."

Karla and Djaloki had me back to my hotel an hour after sunrise. After a quick shower and breakfast I grabbed my gear and Karla and I headed out to the polling places. The lines at most stations were thin or nonexistent. People were apparently afraid of drive-by shootings or pipe bombs, or they were siding with Aristide's political opponents, who'd called for a boycott of the election. We spent the day interviewing those who did come to the polls, nearly all of them Aristide supporters. By late afternoon we were making our way back to the Montana Hotel through very quiet streets.

On the radio reporters in different parts of the country were giving updates on voter participation. Haitian media put the overall turnout at about 10 percent. The Electoral Council would peg it at more than 60 percent.

Whatever the exact numbers, the day had been quiet both in terms of electoral participation and violence. Then on the car radio came the news that Haitians had been dreading. A large unexploded pipe bomb had been found in the courtyard of the Montana Hotel. My hotel. The police were evacuating everyone.

"Translate everything they say," I told Karla. I rolled down my window and held my microphone high in the air as we neared the hotel. If the bomb went off I was going to record the sound of it. We raced up the hotel's long entrance ramp but found the entry clogged with cars trying to escape. I jumped down and ran the rest of the way up the slope. Hundreds of people were gathered at the far end of the swimming pool terrace. In the parking lot, about two hundred yards away, the police had cordoned off a grassy area. Virtually all of the foreign press was there. I saw Trenton Daniel, a wiry, bespectacled reporter from Reuters, and asked him what was going on.

"We're waiting for the bomb squad," he said. "This is a major drag. I had an entirely different story written and ready to file."

Soon a couple of black pickup trucks arrived. Riding on the back of one was a young man in a puffy white jumpsuit whose sections made him look like the Michelin Man. He had on black boots and he appeared as nervous as last night's goat. A man I took to be his boss handed him a tall plexi- glass shield, the kind riot police use to protect themselves from thrown fruit and gasoline bombs. The young sapper pulled a helmet down over his head. I crouched down and moved to a position directly behind him, my microphone pointing directly at him and beyond, toward the bomb some hundred yards away. Trenton had followed me.

"What does the bomb look like?" I whispered. "Did you see it?"

"Yeah," he whispered back. "I reported it. It's a long piece of pipe wrapped in silver duct tape."

"Jeez," I said. "These bastards . . ." Then my stomach dropped. I felt with my hand underneath my gear bag, too afraid to look with my own eyes. Then I forced myself to look. My hands had not lied. I'd just become the day's story.

I stood and approached the sapper's boss.

"It's not a bomb," I said. Now I knew how the late Seattle artist Jason Sprinkle must have felt, with his misunderstood heart sculpture that caused such commotion. For a moment it occurred to me to cut and run like him. But where to?

"Ki?"

"You're not dealing with a bomb," I said.

The policeman yelled out something in Kreyol and an- other officer approached.

"What is this?" said the second man, in English.

"I think your bomb may be my microphone case," I said weakly. "From the description of it."

The man explained what I'd said to the boss. The boss looked at me and spoke again.

"He says you should follow him toward the bomb to confirm," said the translator, pointing to the sapper.

"It's not a bomb," I said.

"He is not sure he should believe you," said the translator.

He grabbed me by the hips and positioned me directly behind the crouching sapper in the puffy whites who began moving cautiously across the parking lot toward the grassy spot where the object in question lay. I followed closely behind, one hand on the sapper's broad and padded back. We were now some fifty yards from the police line and I looked back. Cameras were popping from within the crowd, the bursts of light like muzzle flashes. As we neared a small tree the sapper stopped and motioned me forward. He passed me his shield and pointed at the tree.

"You want me to go over there?" I hissed. He pointed again and took a couple of steps backward. He was retreating. Colleagues of mine in the crowd wore quizzical looks. What the hell was Hadden doing? I'm getting myself arrested, I thought miserably, as I slowly advanced. I had seen Haitian jails.

Then I saw the object. It was leaning nearly upright against the small tree. Indeed it was not a pipe wrapped in duct tape. It was a cardboard tube wrapped in duct tape cut to size to hold a shotgun microphone.

I figured that it must have slipped from its fastenings beneath my gear bag as we'd crossed to Karla's car that morning. I lowered the shield and walked over and picked it up. The crowd, still in the dark, took a collective step backward.

"Not a bomb," I yelled in a neighborly tone. "Just a microphone. Nothing to see here, folks."

Some reporters laughed, but then the local and foreign photographers advanced on me, followed by several angry-looking policemen. I can't tell you how much journalists hate becoming the story. Especially when it's because they've screwed up.

I thought, I'm going to be humiliated, reprimanded, castigated, jailed, fined, expelled. And worse, it's going to be in the newspapers stateside tomorrow.

A policeman I hadn't seen came and stood next to me. He motioned for the crowd to be quiet.

"I would like to say something," he said in a loud, stern voice, in English, "on this day that is so important to Haitians." He grabbed my collar with a gruff hand. "I would like to thank this young man for clearing up this mystery and allowing us to get back to work. There is no bomb. Today has been a peaceful day, thanks in part to the foreign media."

He then shook my hand and mugged for the phalanx of photographers. What I'd expected to be a public flailing now felt like an awards ceremony. I scanned the crowd for the U.S. photographers. As soon as the policeman let go of my hand I jumped into the crowd.

"Hey, you're not gonna send these photos to your editor, are you?" I asked shooter after shooter. "Come on, guys, give me a break here. This is a nonstory."

"Hey, Gerry!" one yelled. "Give us a smile!"

"Hold the mic up higher!"

"Pretend like you're lighting a fuse!"

I implored them not to publish the pictures even as their digital machines were recording dozens of images per second. I was imagining Jenkins opening the *Times* or the *Post* the following morning and seeing my face under a headline about Haiti. It was reasonable to think he would not be pleased.

The photographers, of course, ignored my pleas. I couldn't blame them. Most were freelancers dying to get anything published. And the only damn election drama that day had been me.

That night on the porch of the Oloffson Hotel I received many shoulder slaps. Leading the laugh brigade was Aristide's foreign press attaché Michelle Karshan, the native New Yorker.

"Hey, Ger," she said gaily, "talk about desperate for a story. The CIA ought to hire you to disrupt other elections!"

It was all pretty funny, I guess, but I was having a hard time laughing because I was worried about the morning papers. I wondered if bringing out what was likely to be Haiti's only bomb squad for a false alarm could get me fired. I slept uneasily, but in the end, as far as I could ever tell, only one photo of me ran, in a Texas newspaper, and no one at NPR in Washington saw it. Or if they did they never said anything. The story of the contested U.S. election race between Al Gore Jr. and George W. Bush had probably gobbled up any column inches that might have otherwise been wasted on my brief career as the Caribbean Bomber.

The next day Haitians poured onto the streets of Port-au-Prince. It was literally one giant running party. As predicted, Jean-Bertrand Aristide had swept the uncontested election and hundreds of thousands of his supporters jogged down city streets, carrying Haitian flags and victory banners, playing *rara*-style music, blowing on long plastic horns. Little matter that the opposition, the OAS, and U.S. government officials were not calling to offer congratulations. Revelers sang clever slogans with beautiful melodies about the return of their beloved "Ti-tide." Tiny Aristide, the little priest with his big bookish glasses and soft voice, would now certainly occupy Haiti's presidential palace for the next five years.

On the streets, happy Haitians piled around reporters to express their joy. Each time I told them I was from the United States they would get even happier.

"You Americans are always lecturing us about democracy!" screamed one woman. "We elected our president in one day. And you still don't have one!"

She was of course referring to the flawed ballot counting in Florida that would delay Bush's official victory for weeks to come. That the world's preeminent democratic superpower—and supposed exporter of democratic principles—could have such a colossal flaw in its electoral system unnerved much of the world. It unnerved people, but in some cases it seemed to tickle them.

"We're going to send election observers to *your* country," said another man in a delighted and aggressive tone, "teach *you* how to do it correctly."

"That's right!"

"Listen up, America! Haiti is trying to teach you something!"

I had to smile at the irony. These guys were right. On this day and for the briefest of periods to follow, by virtue of the "dangling chads," those tabs punched but not fallen from Florida ballots, Haitians got to shove democracy in our faces.

The spot where the dirt had been disturbed looked to be about the size of a shipping container.

"That's exactly what it was. Buried right there. A shipping container, filled with weapons."

I was standing on a rickety wooden bridge in a section of jungle connecting Panama to Costa Rica, interviewing the regional head of Panamanian customs. His name was Foster Weeden and he'd driven me out here in his car to show me this.

"What exactly was in the container?" I asked.

"Rifles mostly," he said. "Bullets. Other munitions. Enough to keep a hundred men fighting for a while." Weeden was tanned and relaxed, and when he spoke he leaned toward the microphone like someone testing a public address system.

"And where were the weapons heading?" I asked.

He frowned. "To Colombia," he said. "Like all the shipments before this. Like all the shipments we don't find."

"To Colombia," I said.

He nodded affirmatively. "I've got just a dozen men to patrol this entire border. We stumbled on this shipment basically by accident. There's no way we can stop the weapons flow through here."

"Sounds like you're up to your neck in problems," I said.

"Us?" he shrugged. "Not really. We're just a transit country," he said. "The ones with the real problem are the Americans. These guns are for fighting the Americans."

On that December day, in the year 2000, the United States was not fighting any war, anywhere in the world, except the war on drugs. And that war was now being ratcheted up a notch. It was called Plan Colombia, a new, multibillion-dollar American aid package designed to help the Colombian government wipe out its cocaine industry. That industry lay largely in the hands of an army of former Marxists turned drug dealers called the FARC, or the Revolutionary Armed Forces of Colombia. FARC rebels controlled nearly half of Colombia's territory.

"We suspect that the FARC needs more firepower to hold off the coming Colombian offensive," Weeden said.

"And where do the weapons come from?"

"From your proxy wars," he said.

"From my proxy wars?"

"From the cold war conflicts of the 1980s in El Salvador, Nicaragua, Honduras. When those wars ended a lot of people buried their guns. Probably they thought one day they'd have to use them again. But now it turns out they're worth more to someone else."

"And where does the FARC take control of the weapons?"

"Just across my other border," Weeden said, "in the Colombian half of the Darien Gap."

The Darien Gap, one hundred miles long and fifty miles wide, is a nearly impenetrable mountain rain forest buffer separating Panama from Colombia. It's a territory of dark forests and swamps, jaguars and poison-dart frogs and insurmountable ridges swarming with insects. There are few villages in the gap and none in its deepest interior. But it is not devoid of people. Small groups of Wounaan, Kuna, and Embera Indians make the coastline and nearby islands their home. The Panamanian police were sending regular if tentative patrols up its perimeter ridges, as was the Colombian military on its side. That's because there were FARC irregulars in there too, sometimes taking a break from the fighting to the south, sometimes picking up shipments of arms and supplies to keep their nonideological struggle going.

I left Mr. Weeden, flew back to Panama City, and called a local radio reporter named Larisa who'd agreed to work as my fixer on a foray into the Darien Gap later that week. I was going to need help navigating my way through a region where skirmishes were not uncommon and where getting kidnapped was a real possibility. But when I reached Larisa on her cell phone she told me the trip was off.

"I can't go," she said.

"What?!" I said. "We planned this last week. Our plane leaves at dawn tomorrow. I already bought your ticket."

"I'm too busy," she said.

"Too busy? Larisa, I'm sorry, but you agreed to work for us. You can't just become 'too busy.' If you pull out now you'll be sending a reporter into a very dangerous region without any protection."

"I don't see how I can protect you," she said.

"You know the place. You know the people. I'm not talking about a bodyguard. I'm talking about having a reporter along with experience in a very troubled area. That's worth a lot more than bodyguards."

She hesitated, then said, "*Lo siento*. I'm sorry, I just can't. I, I . . . I can't get the time off."

"But you're freelance!" I screamed. "Listen, I'm going in, with or without you. But if you bail now and something happens to me this is going to be on your head."

"Come back and go a different time," she said. It sounded like a plea.

"Larisa! There is no different time! I'm deep into this story and it's happening now. Please. I'm sorry I'm yelling. I can understand why you're getting cold feet. The Darien Gap isn't exactly the quietest place on earth these days. But we won't stray across the border. I promise."

"Maybe I can find someone else to accompany you," she suggested, talking as if she hadn't heard me. "I will make some calls and call you back. I promise."

The afternoon wore on but Larisa did not call back. That evening her cell phone was disconnected. After dinner I packed my gear, laid out my clothes, and hit the sack. But I was consumed with frustration and some panic and couldn't fall asleep. I stared at the ceiling of my hotel room. That ceiling indistinguishable from all the dozens of other hotel room ceilings I was growing used to gazing at alone in a lengthening succession of nights.

At six thirty the next morning I was standing in line to board my flight to Puerto Obaldia, in Panama's northeastern corner, the last semblance of a town in that sector of the Darien expanse. There I hoped to hire a boat to take me the roughly ten miles to the Colombian border along the Atlantic coast. What I was really hoping was that Larisa would show up at the airport now. The area I was flying into, a vast reserve inhabited by Kuna Indians, had seen exchanges of gunfire recently between Panamanian police and the FARC. It was irresponsible of me to head there alone. But there was no turning back.

I handed over my passport to the flight attendant at the check-in counter. And then Larisa did show up. She was scowling. She punched me in the arm.

"I'm going to kill you," she said.

I hugged her. "I didn't think you'd be here!"

"If you hadn't made me feel so guilty," she said. Larisa was all of twenty-five years old and she was dressed like someone with only the vaguest notion of the outdoors. She'd chosen for our jungle expedition a pair of very tight blue jeans, a tight-fitting T-shirt, and canvas sneakers. She was carrying a tiny knapsack that I doubted contained raingear or bug spray. She was a lovely sight to see. I hugged her again.

"I can't believe it! Sorry I manipulated you like that. Things are going to work out just fine."

"If they don't, I kill you," she said.

"Deal."

After a quick, bumpy flight we took a car from Puerto Obaldia to a small fishing village on the jungle-covered coast. The ocean was calm, the wind still down.

"Who should we ask to take us?" I said, pointing to a group of fishermen on the beach.

"We're going by boat?" Larisa said. She wandered off toward the men, leaving me standing on the jungle's edge with her pack and mine.

"Well, you tell me!" I shouted after her. "You're the guide!"

Larisa talked to the men for a moment. They looked my way. She waved me over.

The boat was old and bulky and wooden, but it had a big diesel engine, strong enough to get us hydroplaning atop the water as we sped toward Colombia. Off the port side the forest swept down from the hills, reaching right to the edge of the high rock cliffs towering above us. The last trees leaned out over the water, bent and swaying in the wind like horses recoiling from a precipice. The sea we were crossing lay in the shadow of those cliffs, and the jungle canopy cast the swell in a rich marble green.

We stepped off the boat at a small, abandoned dock giving way to a level stretch of jungle. Larisa, whose skin was normally a deep shade of caramel, was looking positively ashen.

"You okay?"

She marched off without answering along a small path that our boat's pilot had indicated to us. Soon we came to a small village of concrete huts. On the eastern outskirts, where the jungle began its slow rise toward the Colombian frontier, heavily armed police loitered around a sandbagged building. This was the last outpost in Panama, an otherwise unimportant speck on a map turned strategically crucial by the criminal activity transiting the region.

The police were expecting us and were friendly. I told Larisa I wanted to walk up to the actual physical border with Colombia, if for nothing else than to just gaze across it. She turned on the charm and scared up a young local man to serve as a guide. He led us along a path through a small meadow, and then we were climbing along a steep switchback trail into the trees. The path diverged often, and I was starting to wonder whether our guide wasn't getting us lost: his attention seemed mostly focused on Larisa's breasts. But moments later we reached the frontier, a high ridge that offered far-reaching views into Colombia and back into Panama. Down below on the Colombian side there stretched a long bowed beach, nearly as white as flour, dotted with thatched umbrellas and tables for tourists. Not a soul moved upon it.

Somewhere, deep in the forests behind, FARC rebels were allegedly ensconced in their hidden camps, waiting for weapons, sleeping off combat fatigue, readying themselves for more bloodshed.

"The tourists no longer come here," offered our guide.

A Panamanian police captain, standing at a nearby hut listening to a radio, laughed at that.

"May I ask you a couple of questions?" I asked.

He looked at my recording device suspiciously. "Only if you turn that off," he said.

"*Ningun problema.*"

And to show him so I unplugged my shotgun microphone from the digital recorder. Then for some reason I did something really stupid. I quietly activated the recorder's built-in microphone. After traveling all this distance, I thought, I just couldn't leave here without an interview, recorded, with somebody in charge. Not only was my trickery ill-advised; it was surely illegal. I immediately regretted it, but once I'd pressed the button there was no way to turn the machine off without calling attention to my sneakiness. I held the recorder nonchalantly in one hand as I asked him about FARC activity in the area.

"They could overrun us at any moment," he told me. "We don't have the men or weapons to fight their army."

"But they wouldn't," I said. "That would almost certainly lead to a major diplomatic incident, if not outright war."

"No, they won't launch a big attack," said the captain. "But they make sure we know they could. As long as they've got us cowering behind sandbags, they can go on with business as usual."

"That business being?"

"Their business? What do you think their business is?"

"You tell me."

"You come all the way here and you don't know the FARC's business?"

"Yes, I do. But can you clarify that?"

As my captain opened his mouth to speak he was interrupted by a sharp click. He looked down at my hand. The one holding the digital recorder. The minidisc inside the recorder had reached its end and the machine had automatically shut off. On that particular model when the record button disengages it makes a very audible clicking sound. I was busted.

"Were you recording me?" he asked.

I held the machine out to him. "No, I wasn't. That was the cover. I accidentally opened it. Then closed it. Take a listen if you'd like."

The captain took the machine and looked at it.

"What's this button?" he said.

"That's pause. To stop playing."

"What's it for?"

"To stop the machine from playing."

He flipped the machine over once, then handed it back to me. "Their business is drugs," he said. "Drugs and guns."

"How do you know?"

"The drugs everyone knows. The guns, hang around for a while." He looked at my recorder again. "Interview over."

That night, as Larisa and I were preparing our cots in a concrete hut back in the village, the captain and his men opened fire on the darkness around them. A dizzying siren went off. Larisa came running into the room from the latrine and crawled under her bed. Outside people were running and shouting. I grabbed my recorder and dashed out. I found a group of police heading for the sandbag positions at the edge of town.

"*Voy con ustedes,*" I said. *I'm going with you guys.*

Back at the hut Larisa asked me where I'd been.

"I wanted to tag along with the police," I said. "But one of them cuffed me and ordered me back to the hut."

"Who's out there?"

"I don't know."

The Panamanian police didn't know either. The gunfire was sporadic and never heavy, but it lasted the good part of an hour.

The next day, the captain explained that they'd seen lights moving on the ridge above the village. That had been enough to get their guns blasting. I now understood his point about being kept cowering behind sandbags. The FARC didn't have to attack Panama's understaffed police outpost to keep its men pinned down. They just needed to turn on a couple of flashlights.

The following morning we met two reporters from Agence France-Presse, the French newswire, who were working on the same story. I told them we were planning to go to some nearby islands to interview Kuna Indians who'd allegedly witnessed strange activities in the area, most likely related to the FARC. They decided to join us.

"We're going by boat again?" Larisa asked.

By boat, but not on a modern craft the likes of which had delivered us to the border. To reach the Kuna communities on the low islands along the Atlantic coast we would hire a Kuna pilot and his less sophisticated, less comfortable vessel. The Kuna themselves motored about the sea in long wooden boats, as narrow as canoes, with small outboard engines screwed to the back well. The steering rudder was attached to a long wooden pole that extended into the water off the stern. The pilot had to stand to steer; this also allowed him to watch out for big waves that might flip his vessel.

Half an hour later we reached our island destination. The Kuna from this particular village were out near the beach reconstructing the roof of their main meeting and dining hall. Men, shirtless and wearing swim trunks, labored alongside their women in blue blouses and red embroidered head scarves. The thatched canopy roof on their long pole house had dried out under the tropical sun and partly collapsed. The Kuna chief, an old man with mirrored sunglasses, said the seven hundred residents would gather as soon as the repairs were finished to discuss the mysterious occurrences.

"We're going to discuss Plan Colombia," he said, "because we're seeing strange things in the area. Planes and helicopters swooping over us. Just two days ago one passed over the river just behind. We assume it is due to Plan Colombia. We have never seen such aircraft before. They are not Panamanian. At night we see flares coming from areas where no one lives."

Abruptly, the chief said he had no more time for such talk and turned his back to us. I imagined that he considered my visit to be just another unpleasant interruption related to the military and paramilitary maneuvering.

"To get back to Panama City is even easier than getting here," an aide to the chief told us. "The quickest way is to continue with your Kuna pilot to a nearby island. In the morning a small plane will land at a jungle airstrip to pick up anyone going to the capital. The timing is good," he said, "because the plane only comes by once a week."

"How far is this other island?" Larisa asked.

"You will follow the coast for a short while and then cross an hour of open sea."

"I'm not going," Larisa said.

"Again you're not coming?"

"I'm not crossing the open sea in one of those boats."

"They are very sturdy," said the chief's aide. "And your pilot is experienced in these waters."

The AFP reporters came over and assured her in their heavily accented Spanish that everything would be okay. Finally, feeling embarrassed and fearing that she would hold up the entire trip, she agreed to one more boat ride.

But she would no longer look at me.

We all climbed aboard at the beach, and the pilot and his adolescent son heaved on either side until the dugout slipped forward and off the sand. We sat idling for long minutes just a few yards from the shoreline, the prow pointed toward the breaking surf some three hundred feet farther out to sea.

"What are you waiting for?" I asked the captain.

"A gap in the waves," he said. Another Kuna boat idling near ours made a run at the surf. When it reached the waves it crested the first safely. But a second incoming curl, dangerously close to the breaking point, lifted the boat almost vertically into the sky. The pilot crouched low into the hull and ducked his head. The prow pierced the top of the wave and plunged forward. The momentum of the rushing water now tossed the stern skyward. For a moment the boat seemed to be moving in reverse. Its engine coughed loudly, sucking in air. Then it was on flat water again and gunning full throttle for the open sea.

"You weren't watching that," I said to Larisa.

"I don't know how to swim," she said.

"You don't know to swim . . ."

Our captain floored his engine. I stuffed my gear bag between my feet and watched the approaching surf, now a blue-gray wall with the wind at its back. We all clung instinctively to the dugout's runners. Like the boat that had gone before us, our prow rose quickly to an angle suitable for launching rockets. Then it came swooping down hard. Not a second later another wave hit us diagonally and filled the bottom of the boat with water. Larisa took the worst of it. She had curled up in the fetal position in the bottom of the boat. She might not have known how to swim, but she was now literally floating in water at our feet. She began to cry.

"*Ya esta*," said the Kuna pilot in his broken Spanish.

It was true—it was over now. We'd broken past the surf. Before us now lay an endless and choppy sea licked frothy by the wind. The pilot produced half of a plastic bottle from who knows where and handed it silently to his young son. The boy asked Larisa if she wouldn't sit up please. Then he began bailing.

For a good hour he bailed. But the water level didn't seem to be going down.

"Everything okay with this boat?" one of the French journalists asked.

"It is true that it is leaking," the pilot said. "We must have cracked the hull breaking through the surf. Do not worry—my son will keep us afloat."

Larisa was shivering uncontrollably now in her seat. Between the wind and the chop we were being constantly doused with cold seawater. One of the French reporters opened his backpack and pulled out oversized plastic garbage bags. We pulled them down over our heads and poked holes in the bottoms to pass our heads through: instant ponchos.

The plastic kept both the water and the wind off, and the ride became more bearable. I asked Larisa where she was from originally.

She did not answer.

An hour later I could see no sign of another island.

"It is a bit farther," the pilot said.

After three hours I caught sight of land off the starboard side.

"Is that our island?" I asked.

"That is the coast," the pilot said. "We have been following it all afternoon."

The hammering of the small waves against the bottom of the boat was beginning to make my back hurt. But I could hardly complain. The pilot's son was still bailing, hunched over, sometimes on one knee, sometimes on both.

At sunset the sea suddenly grew calm, nearly glasslike but rolling slightly, as if all the day's fuss and bluster had been about nothing. Now we were slicing through the water at a better speed and the pilot said that we were close. We'd been in his boat six hours. Larisa was curled up in the boat's well, half soaked but sleeping somehow. I shook her awake.

"Get up. Get up," I whispered.

"*Por Dios.* What's wrong?"

"Look."

She pushed herself into a sitting position and peered over the side.

"*De donde vinierion ellos?*" she said. *Where did they come from?*

"Shhhhh," hushed a Frenchman. "Shhhh."

There were three of them swimming alongside us, maintaining exactly the same speed, their beaks just below the surface of the water. Their skin like latex, impossibly smooth. And it seemed that for all their speed they were making no particular effort. They breached the surface, stones skipped perfectly. Above the putter of the engine, only a faint hiss as they cut back into the water. Over and over again.

I no longer felt cold or tired. My recording gear lay useless on my lap, secure inside my bag wrapped in plastic against the sea. The sky reddened and still they followed. Larisa was leaning over the edge of the boat to watch one crisscrossing beneath us. I held her plastic poncho with one hand just in case.

"They teaching you to swim?" I said.

She held a finger to her mouth. Her expression: I am not afraid. In the distance an island appeared like a low gray cloud, and soon I could see fires burning and lights.

When I looked for the dolphins again, they were gone.

At the dock our pilot and I helped Larisa get off the boat. She lay on the wooden planks for a long minute before standing up on shaky legs. We were all a bit wobbly. We entered a small Kuna village consisting of wooden huts and a bar. As we passed the huts you could make out people moving inside through the vertical slat siding. The pilot deposited us at the bar. We invited him for a beer, but he waved us off and disappeared, his son in tow.

"Hey, we made it." I raised my beer.

"I still get to kill you."

That night the Kuna let us sleep in hammocks in their

longhouse. We went to sleep in our wet clothes. The next morning we trekked out some two miles through the jungle until we reached a very short and bumpy airstrip. A small eight-seat plane was already there, its engines running. We shoved our gear in its aft storage compartment and climbed aboard. We handed the pilot a wad of cash, and he turned the plane around and accelerated.

"Tonight's my last night in Panama!" I yelled to Larisa as we bumped along. "I'd like to take you to dinner! To make up for all this!"

"I don't think that would be a good idea!" she yelled back.

"What? I can't hear you! The engines!"

But she shook her head from side to side to clear up any possibility of confusion.

When the plane reached the end of the runway it fell toward the sea below and used that momentum to swoop back upward. The waves were a hundred feet below us. The same sea we'd crossed the day before. The same sea the gun-runners were using to move their contraband south toward Colombia. The Indians had seen them at night. Panama's police were seeing lights. The lights of a shadow army tens of thousands strong, occupying half a nation, exporting cocaine to the whole world.

And the United States was about to fund the largest military effort ever to stop this. In the war on drugs the Americans would literally come up against some of the same guns they'd paid for—or fought against—in Central America in decades past. Same defects, same serial numbers, different fingers on the triggers. Then, the enemies were supposedly ideological. Now, they were well-armed businessmen running a multibillion-dollar cartel that neither pledged allegiance nor paid tribute to any state. As guns moved toward Colombia over the next nine months so did hundreds of millions in U.S. dollars designed to counter them. The way things were shaping up, it seemed I was about to become a war correspondent.

The vast field of mud I was standing on had formed part of a Salvadoran mountainside just hours earlier. But an earthquake had shaken the region for thirty seconds and the mountainside had collapsed. A middle-class neighborhood, home to some two thousand people, was buried almost instantly under a crashing, twenty-foot-high wall of dirt and debris. Aghast witnesses described the landslide as like a wave, with trees and bricks and cars and dogs and telephone poles tossed in a flotsam and jetsam of roaring brown earth that rushed at them faster than anything they had ever seen.

From a distance, the disaster site looked like a coffee spill against a green tablecloth of Central American forest. Now I was standing on top of it, amid the mechanical coughing of yellow construction backhoes, the shouts of Mexican rescue workers with their world-famous sniffer dogs searching for people dead or alive, the occasional reporter slipping up to his knees in muck, bewildered police and soldiers, red-faced

Red Cross workers, and, above all, neighbors, children, fathers, friends, grandmothers, in-laws, and strangers clawing with their hands and with sticks in spots where they could only guess their houses had once stood.

The paralyzing din of machines. The frantic cries. Emergency lights of every color flashed everywhere. At some point in that cacophony of panic, as I was running about trying to find somebody, anybody, who could explain to me the extent of the destruction, I noticed several mud-covered rescue workers in one area atop the spill urging each other to shut up, their fingers pressed against their lips. They stood out precisely because they were standing so still. I picked my way over to them through the soft mud and debris. The going was slow because I couldn't be sure what was beneath me, and there was constant danger of falling through the mud into some empty, hidden cavern below. When I reached them I saw in their midst a young man in a light blue button-down dress shirt and khaki pants. He could have passed for a Microsoft employee on casual Friday except that his pants were caked with earth from the thighs down and his face was wrenched with worry. He raised a black cell phone to his ear. Shhhhhhh, urged the men and women around him in that chaotic manner that creates more noise than the noise you're trying to suppress. But finally most everyone near the man grew silent. And then I heard what everyone was listening for: a faint ring. It sounded three times, then stopped.

"Who are you calling?" I asked quietly after the young man had hung up.

"My fiancée," he said. He said his name was Ignacio and he pointed at the earth with a pink trembling finger. "She's underneath us."

Across the plain a group of rescue workers began to yell, "Body! Body!" One of them emerged from a muddy depression with a shorn torso and began jogging it like a heavy log toward the periphery of the mud. A woman waiting there fainted.

"It just rings three times, then goes dead," Ignacio said. "She dialed our local emergency number just after the earthquake," he said. Then he added hopefully, "Maybe reception has been cut?" He said it more like a question. He raised his phone again and everyone shushed each other, and when the faint ring started they threw themselves at the sodden earth and dug.

The quake's epicenter was located some sixty miles southwest of San Salvador beneath the Pacific Ocean, and it shook buildings as far away as Mexico City, though when it struck that Saturday morning I did not feel it. I was sitting at a café near the bureau in the Distrito Federal reading about the slowing U.S. economy and its potential effect on Latin America, especially Mexico.

As I was reading my phone rang. It was an NPR weekend producer in Washington calling to ask if I'd felt something.

Back at the office, I found that all flights into San Salvador were canceled. The international airport's runway had been cracked to pieces by the quake. So I booked two seats to Guatemala City, the nearest major hub, and called my colleague and friend, Bernadette Rivero, a young genius with a knack for offbeat stories who was stringing for the Weather Channel. I figured she could earn a few extra bucks.

"Earthquakes count as weather, right?"

"Not really," Bernadette said, "But I think they'll bend the rules." Bernadette was the only person I'd ever met who could speed-read. It took her ten seconds to read two pages of a book. I knew she wasn't bluffing because I'd tested her with random volumes off the shelves in the office library.

We landed in Guatemala City, jumped in a rental car, and sped south along the winding Pan-American Highway. As we approached the Salvadoran border Bernadette pulled out a tube of lipstick and wrote PRENSA in big red letters on the windshield. At the border checkpoint Salvadoran customs officials saw the press sign and waved us in frantically—they were anxious for word of the disaster to reach the outside world, for the slow engines of international aid to begin firing. Soon we saw why.

At the regional capital of Santa Ana, about two hours up the road from San Salvador, we jumped off the highway after spotting some emergency vehicles. We followed them to the town's main plaza, a modest square surrounded by forest and host to a massive stone church built in the late nineteenth century. Exactly half of it had been reduced to rubble in the quake. The other half stood undamaged. We parked and ran across the courtyard just as a rescue crew was dragging out the church's only victim, a young woman who'd been working alone in an upstairs office when the earth heaved. Her dark skin and tightly coiled hair were caked gray with cement dust and her face was crushed indistinguishable. The men lay her down on the dirt by the main gate, and one of them pulled her mussed skirt down over her thighs as a crowd of locals drew close.

"*Es un milagro que hubiera una sola persona dentro del templo,*" an old man commented. *Only one person was inside the church—a miracle.* Others nodded in agreement. People began filing past the body and gently dropping coins and paper bills onto her stomach, chest, and neck. I took it for some superstitious act until one resident told me the gesture was practical.

"The money will help pay for our neighbor's casket and burial."

Then we were racing toward the capital again. The going was fast once we reached San Salvador as hardly anyone dared take to the roads. As we passed poor neighborhoods of low, unpainted cement houses I saw surprisingly little structural damage, save for the occasional partly collapsed wall. But reporters on the radio were talking about a big landslide in Las Colinas, a mountainous area just on the edge of the city. We made our way there and that's when we first saw the collapsed mountain and the plain of mud lying atop the neighborhood it had covered.

Bernadette and I separated during those first few hectic hours. Later I found her filming the chaos from a muddy knoll, and we made our way to the car and then to the Sheraton Hotel, which was apparently still structurally sound. I was in my room, putting the final touches on my first story over the phone with Glickman, when an aftershock hit, then another.

"What's that noise?" Glickman asked.

"The tiles in my bathroom falling and shattering," I said.

The third temblor was strong enough to convince me to drop the phone and bolt out my hotel room door. I was aiming for the emergency stairs at the far end of the hall. But just one step outside my room I found myself face to ashen face with a man who had evidently held it together enough to do what you're supposed to do during an earthquake: place yourself in the archway of a door where structural support is theoretically strongest.

"You'll never make it," the man said calmly as I ran past. I looked back. For a guy who had just offered me his opinion in such a measured tone his knuckles were an awful shade of purple where he was gripping the doorframe. The aftershock stopped. I stopped. "If this hotel goes down," he said, "the stairs won't help you."

"Silly me," I said.

I reached out my hand and introduced myself. The man let go of his doorframe.

"Kevin Sullivan," he said. "*Washington Post.*"

"And that doorway's gonna save you?" I asked.

"I'd rather die in a doorway than get crushed on a staircase."

Without further debate we went back to our respective rooms to continue filing. Then I saw out my window that many reporters had already set up shop outside, around the pool or in the open courtyard where they were out of any crumbling building's reach. I grabbed my gear and satellite phone, knocked on Sullivan's door, then headed outside to the relative safety of the open air. And that's where Lazaro first appeared.

In those tense hours after the quake, on January 13, 2001, a TV producer for the Associated Press was hustling with the rest of us to set up her workstation on the Sheraton's patio. She was in the middle of chewing out a production assistant for not having packed a critical cable when she felt a tap on her shoulder.

"Excuse me," a dark-skinned, curly-haired young man said to her. "What kind of cable do you need?"

"XLR to mini," she said.

"Not a problem," the young man said. "I'll be back in fifteen minutes." The producer promptly forgot about him. Fifteen minutes later he returned. With the cable.

"I don't know where you're from," she said, "but from this moment on you stick by me." For the rest of the week Lazaro Roque worked for the Associated Press Television News, doing everything from keeping track of cables to scrambling up lunch for the crew.

What made Lazaro's small contribution to that day's

news so remarkable was that he wasn't from the capital. He wasn't even from El Salvador. He was from Guatemala City. He'd traveled six hours by bus to the disaster site, with no contacts and no connections, in the hope of finding some work. It was a lark, a desperate lark, and it paid off. Soon he became one of the foreign press's most coveted fixers in Guatemala. And that's how I came to hear of him.

Lazaro and I were racing across the Petén, a vast stretch of grasslands extending from north of Guatemala City to the border with Mexico, in a rental car meant for better roads and slower speeds. We had a lot of ground to cover. We were searching the northern corner of the region for a camp of Guatemalan refugees. They were refugees who'd returned from Mexico following the end of Guatemala's civil war in the mid-1990s. The exiles, mostly poor Mayan farmers, had been lured back by government promises of amnesty, land, and economic support. Instead they were dumped in remote, impromptu villages that lacked even the most basic services such as running water, doctors, schools, or telephones. These settlements were reachable only by dirt tracks that became impassable with the rains that fell during winter. Several years had passed since the government created these camps, and it had done nothing to improve the quality of life there.

This was the first time Lazaro and I worked together.

"There's a map in the glove compartment," I'd said, as we'd pulled out of my Guatemala City hotel parking lot earlier that day.

"I don't need one," he said.

"You've gone there before?"

"Not exactly. I looked at the map the other day."

"What, you memorized it? Do you realize what we're looking for? We could get lost."

"*Poco probable,*" Lazaro said. *Slim chance.*

Six hours later, when we finally left the pavement behind and went crashing along through a maze of deep, rutted tracks overgrown with wild grass, he still seemed to know which way to go. "Left here," he said, as we approached one barely visible fork in the trail. Then, an hour later, "Now up that slight rise. Left again. Right."

"How the hell do you know this?" I ask.

"It's gotta be this way," he laughed. "Trust me."

"Holy shit!" I yelled, swerving hard into the grass. A pickup truck had appeared out of nowhere, speeding at us from the opposite direction. Seated in its open bed were three beefy, serious-looking men.

They spun past us and I was able to steer us back onto the dirt track.

Lazaro said something under his breath. I saw him glance several times in the sideview mirror.

Then we rounded a bend and found ourselves right in the center of one of our lost camps. Lazaro smiled triumphantly.

"You've been here before," I said.

"No, I haven't."

The resettlement "village" consisted of a dozen wooden huts with corrugated metal roofs and a larger, central building that served as a meeting hall and collective shed. A loudspeaker was strung up on a lone, tall pole. The Indian villagers were not expecting us but then again without phones I assumed they rarely had a heads-up about visitors.

Their leader led us into the meeting hall and offered us wooden chairs. He was a deeply wrinkled man in a crumpled straw peasant hat.

"Government help is not going to arrive any time soon," he said. He spoke in slow clips, isolating each syllable. He was Mayan, and Spanish was his second language. "But we're not rolling over either. We grow what corn we can. After harvest we usually walk it out several miles to the nearest paved road, then flag down passing vehicles to get the corn to market.

"We've also built a schoolhouse for our children. They have no books, but they have classes."

When at midday the pickup truck that had nearly careened into us suddenly rolled into camp, the faces of the villagers watching the interview turned to stone. The leader stood and got on his rudimentary public address system and began calling all the men in from the fields to the meeting hall. The four men from the pickup truck dismounted their diesel horse and sauntered over to us like extras on the set of *Bonanza*.

"How are you today, sirs?" said one, smiling deferentially.

"How is everything going in the village?" The villagers regarded them without expression. Several people came running in from the fields, machetes in hand. I knew something was going to happen then, but I didn't know what.

"If you would be so kind," the visitor continued, "we have come with a message. Our *patron* is planning to reclaim some of his land here. He has no choice. He's been cheated, you see, abused by the authorities, as scandalous as that might sound. They've given to you what they stole from him. You will need to move the official boundary markers of your land closer together, cinch your belts a bit, so to speak. Surely you'll help our *patron* correct this grave injustice, no?"

The village leader looked up and smiled from under the brim of his hat.

"No one will be touching any property lines."

"Listen, *viejo*," the visitor said. His tone changed, but his smile remained, like a mask that no longer fit right. "You'd do better to understand our boss's position. He was here before you."

"My people have always been here," said the village leader. "Let's go swimming."

We followed the old man down a long path through the high grass. The thugs, momentarily thrown off guard, tagged along. We came to a creek lined with trees, and everyone stripped down to their underwear and jumped into the cool water. Surprisingly, the goons jumped in too. They kept to themselves but managed to bother everybody with their brash laughter and deliberate splashing.

"These fools better wise up," said one them out loud. "They don't know who they're messing with. I'm sure they don't want their troubles to resume." But the Mayan leader just floated on his back in the sunshine. Without his hat on he looked much older and more vulnerable—like some wizened turtle out of its shell. But he didn't take the goons' bait. Eventually they climbed out of the water and got dressed.

"We'll be back soon to hear your answer," one of them said. "And it better be different."

"*Puchica*," Lazaro said to me when they were gone. "That was close. Things would have gotten a lot more tense if we hadn't been here."

We got dressed ourselves and returned to the village shed.

"Take my card," I said to the leader. "Call me if you should have any more trouble from those *matones*." He nodded, but I knew he would never call. This old Guatemalan man had no idea who I was, and given his circumstances, he had just as much reason to mistrust me as anyone else. He tucked my business card into his shirt pocket in a gesture of courtesy.

We spent the night an hour away in a small terraced hostel on an island in the center of Lake Petén Itzá, a beautiful freshwater body in the center of the Petén. Lazaro and I sat on the roof with cheese sandwiches and cold beers as the sun set and the mosquitoes rose.

"Man, I can't wait to get home to my girl," Lazaro said.

"You've got a girl?"

"Yes. Susana. She's a freelance TV reporter. I think we're going to get married."

"You did good today," I said, changing the conversation. "I'm going to expose your government's negligence now that we have indisputable proof of it."

"I hope what you write makes a difference," Lazaro said.

"Either way," I said, switching to English, "you are one bad-ass tracker." Lazaro watched me, half smiling, waiting for the translation.

"I don't know how to say 'bad-ass tracker' in Spanish," I said. "But I never would have found that settlement without you."

He raised his beer can in the air and grinned broadly.

Then he turned serious. "My government is holding Guatemala hostage," he said. "We live with the same feudal structures the Spanish set up here five hundred years ago. Nothing has changed. We live like slaves."

"But young people like you seem to be trying to change things," I said.

"*Matan a jovenes como yo todos los días aca.*" *Young people like me get killed every day here.*

"And you? You seem to be doing okay."

Lazaro threw me a beer and opened another. "In Guatemala the earth can open under your feet at any moment," he said. "It's opened under mine too many times to recount."

"The most recent, then," I said.

"Okay. Two years ago, in the first year of my university studies, I bought this thirty-five-millimeter camera. One day I was walking downtown and I saw the police beating a man. I snapped a frame and got thrown in jail for my trouble. I most likely would have rotted there, but my good grades saved me. At the time I was at the top of my class.

"After a few days one of my professors noticed that I was absent from class and he began poking around. I was crammed into a tiny cell on the other side of the city with a bunch of young guys who'd been waiting years to be charged with any crime. One morning they announced that they were going to let one guy go. I slipped him a note and he managed to smuggle it out for me. When my *profe* got the message he called the university rector, who lobbied for my release. Probably they paid a bribe. I got out in time to take my exams that year."

"And your family? They couldn't help you?"

Lazaro's face darkened. "My father isn't the helpful type. I spent my childhood traveling around Central America watching my papa wrest change from people poorer even than us. He was a carnival ride operator. Mainly he's just mean. Mean enough to drive my mother away. Mean enough to drive one of my two older brothers away."

"And you?"

"He drove me away too—straight into books. School was my thing. Until last year when I had to drop out to take care of my mother. I always paid for my studies. I used to put on puppet shows for gringos like you!" he laughed. "Right downtown in the old colonial center. That was good money."

"That's a dangerous part of town."

"I am not afraid of anyone. That's enough."

We fell silent again. It grew darker. The mosquitoes were devouring us and we stood to go inside.

"One bad what?" Lazaro said.

"One bad-ass tra-cker," I repeated, isolating each syllable. And in what remained of the light I could see Lazaro mulling over the phrase. He seemed pleased with the ring of it.

The entire town was here to dress me. At least that's what the sign said: *Estamos aqui para vestirles.* It was written on a big plaque at the base of a giant statue of three spools of thread, blue, white, and red. This is what welcomed you into the Mexican town of Uriangato, not far from the ranch of Mexico's president, Vicente Fox. I was here to talk to residents who regularly immigrated to the United States. Because their trip was about to get a lot easier.

Fox's democratic victory was finally bearing fruit. After months of high-level negotiations with Mexico, President Bush let it be known in the spring of 2001 that he was making a new guest-worker program a top priority. The blueprint under discussion would create three-year work visas. Migrants would be allowed to bring their spouses. If the proposal became law, it would transform the way Mexicans crossed the border. That is, it would deal a major blow to the human smuggling rackets operating from San Diego to Brownsville.

I drove past the Welcome Spools toward Uriangato's main square, passing storefront after storefront of spooled cloth and ready-to-wear T-shirts, bras, and jeans displayed by the thousands on plastic hangers. Rolls of textiles six feet tall leaned up against stucco warehouses.

Suddenly somebody on a shiny Vespa scooter shot past me, beeping. Then another. Then a whole group of them, like Mexican Mods minus the gabardine. The scooters they rode were top-of-the-line models decked out with high windscreens and chrome and dual sideview mirrors. I thought, maybe there's a rally, like a Harley-Davidson gathering. I parked my car alongside a small and exquisite central square ringed by smart houses and a pillared, arched promenade to keep the sun off storefronts and stoops. It was evening now and hundreds of starlings crisscrossed the sky, feeding on clouds of gnats. The plaza was tightly planted with orange trees grown to some ten feet then pruned to allow their canopies to intertwine. In the warm air, *ranchera* music and the Doppler drone of the scooters. I approached a chubby young man sitting on a scooter licking an ice-cream cone.

"What flavor?"

"What?"

"The ice cream."

"Who are you?"

"A journalist."

"A journalist from where?"

"From the States."

"Are you writing about ice cream?"

"No. About people who go up north to work."

"Then you've come to the wrong town," he said.

"I know," I said. "Which is why I'm here."

"Your Spanish is good. But I don't understand you."

"Could I ask you a couple of questions?" I said, pulling out my recorder.

"If you'd like."

"What's your name?"

"Francisco Tonorio."

"How old are you?"

"Twenty-six."

"You'd seem to fit the classic profile of the young Mexicans who strike out for the U.S."

"I have never been to the U.S."

"Never?"

"People go because they need to go. They do not go for the pleasure of leaving home."

"How'd you afford the new Vespa?"

"Working," he said proudly. "How else? I'm a truck driver. My salary is very good, compared to other places in Mexico. Uriangato is industrious. There's tons of work here. And usually it's well paid."

"That's extraordinary," I said.

"Go talk to the mayor," Francisco said, finishing his ice cream and putting his helmet back on. "The mayor can tell you all about it. He's the reason for all of this."

I swung by the mayor's office. Twenty years earlier, Carlos

Guzman told me, he'd left Uriangato to seek his fortunes in the United States.

"For many years I worked mainly as a gardener in California," he said. Guzman was wide-faced and forty-two years old, dressed in a Polo-style shirt and flip-flops. His office had floor-to-ceiling windows overlooking the square.

"I never became rich, at least not by American standards. But the money I was able to send back to Mexico went a long way—and not just for my family. I gave some of my earnings directly to the town itself."

"That's one way to become mayor," I said.

"We built the square. We sent communication equipment to the police. We sent money to build part of the church. Over the years I also collected tens of thousands of dollars from other *Uriangatanos* living in the U.S. When I first came back to town, to oversee the renovation projects, my neighbors here didn't know how to react."

"But over time the idea caught on."

"Eventually, yes. When people saw the changes they sent even more money from the United States. Many began to invest their dollars in local businesses, mainly in textile factories and supporting ventures that could operate independently from the seed money wired from America.

"Over time our collective wealth has grown," Guzman said. "Now we've reached an astounding milestone: virtually no one leaves for the U.S. anymore. They don't need to."

Guzman had to tell his story somewhat loudly; outside, just below his windows, the youth of Uriangato were revving around the main square on their fancy scooters. It seemed like a luxurious amount of horsepower to waste driving in circles, but it underscored his point.

The next morning, two miles away, just beyond earshot of the motorbikes, another world. A dirt road pitted and lined with debris. Plastic bags tangled on brush field upon field. The track into town trodden more by those leaving than those coming home. El Charco, population twelve hundred. And barely a man among them.

I rolled along the dirt track into the village until it ended in front of a small cement depository. No plaza here, no ice-cream stands, no signs of investment. Just squalid huts of mortar. A few fruit trees. The elderly in doorways, watching.

Before me a woman sat crouching beside a low stone wall, her back to a field of maize. The swaying corn a soft dry rustle. The car engine ticked as it cooled.

"What are you doing here?"

"I am waiting."

"For what, *señora?*"

"For my husband. It is not funny."

"No, no, it is not." I hadn't laughed. "What is not?"

"I have four children to feed. How am I going to do that? We have only maize and peppers to eat. Whatever we can grow."

"When will your husband come back?"

"The men here have all gone away."

"To the north?"

"My husband has probably remarried. We have had no word. He sends no money. We have nothing."

"How long has it been?"

"Nine years."

Her eyes wholly unlit, her skin a farmer's, sun-toughened, creased. Just crouching by a low, stone wall as if taking cover from gunfire. An old man in a brown coat shuffled past. "*Buenos días*," he said but did not reach for his hat.

"What do you think you will do, *señora?*"

"If I could, I would go north as well. But with what money? We are trapped here. None of the men has sent word. *Hijos de la chingada*. Sons of bitches. I want to leave too, before my husband returns like him, *gastado*. Spent."

The old man in the brown coat, leaning his head now against a wooden barn, talking softly, hat in hand. Words perhaps for a long-departed son or for a woman he himself once knew in the north. Or simply for the chickens inside. . . .

The doorways, all empty now. I walked the perimeter of the village, crossing no one's path. In a far field two women were planting or weeding. I approached, but they waved me off. With no other possibilities I went back to my car and drove away. Sure as a native son that I would never return.

Berlin, El Salvador. Coffee plantation foreman Rafael Castellano has no workers to oversee. They've all gone to America.

Migrants from Central America rest at a shelter before the most dangerous leg of their trip to the United States: crossing the Suchiate River from Guatemala into Mexico.

For about fifty cents, a poleman takes migrants across the river from Guatemala into Mexico. He doesn't ask for ID.

Contraband, human and otherwise, crosses the Suchiate River in plain sight of Mexican and Guatemalan customs officials.

A Salvadoran man with a broken back and maimed hand convalesces at a shelter for migrants injured trying to ride the Mexican "death trains" up to the U.S. border.

A moveable watchtower keeps an electronic eye on the U.S.-Mexico border near Sasabe, Mexico.

Outside Tucson, Arizona, rescue workers tend to victims of a crash. Twenty-three migrants were hidden in the van when one of its tires blew.

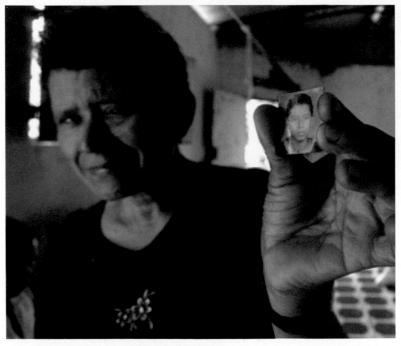

Santa Clara, El Salvador. Blanca Romero holds up a photo of her son who died crossing into the United States. She was told that his heart gave out under the Arizona sun. He was twenty-five.

A hillside neighborhood in Port-au-Prince. Poverty and shoddy construction caused houses like these to collapse in the 2010 earthquake.

Aristide supporters march through Port-au-Prince demanding that their president be allowed to serve out his five-year term.

A young woman helps a schoolgirl who fainted during an anti-Aristide protest in Port-au-Prince, days before the president fled the country.

Haitian "rebels" and members of the Cannibal Army march through Port-au-Prince on March 1, 2004, the day after Aristide fled the country.

Into the chaotic atmosphere rolls a Christian missionary named K. A. Paul. He stages a food giveaway and causes a riot as desperate Haitians fight over his meager offerings.

James Petit-Frère, a.k.a. Bily, or Bily Iron Pants. From Aristide orphan to powerful gang leader in Port-au-Prince's largest slum, Cité Soleil.

Haitian kids watch U.S. Marines take the National Palace in Port-au-Prince, beating the rebels to the prize.

If President Fox could secure a guest-worker deal, villages like El Charco might well be saved from their misery and oblivion. To that end, to further entice his American counterparts, Fox flew to Washington with a gift for President Bush: he was going to seal off Mexico's porous southern border. By doing so Fox would effectively block all other undocumented Latinos—from Guatemala and points south—from reaching the U.S. line. Mexico's neighbors cried foul, but Fox brushed them off. He was too close to clinching a deal to let anything—even Latin American solidarity—get in the way.

That first week of September, as Fox was getting the red-carpet treatment at the White House, I was gearing up for a trip to Mexico's southern border. I was going to cross it illegally myself, to demonstrate just how easy it was. But then I was told to postpone. Instead I was to be dispatched posthaste to Bogotá, Colombia. Plan Colombia was about to enter an even more aggressive phase in its press against the region's coca growers and cocaine producers, and the United States was poised to announce a significant boost in funding. U.S. Secretary of State Colin Powell was already en route to the region to deliver billions more dollars in aid and guns.

I went to bed early the night before my flight. At some point I was awoken. I don't know what time it was because while it lasted, and until I fell back asleep, I was facing away from my clock and did not lift my head from my pillow.

I was lying flat on my stomach, eyes open. It was not a dream and I am my own witness. The feeling I had was that someone was in the room with me. I rolled my eyes, glancing up over my shoulder, and saw a dark sort of patch hovering there, different from the normal night. It was more than just an absence of light. I think now that it was the absence of even the chance of it. The familiar fear and weightlessness that always accompanied these unexplained events in my unexplained house rose in my gut. Equal parts excitement and dread. I waited motionless.

Then I felt a slight pressure on my back. At first it was so faint I thought my mind was playing tricks on me. But it gradually increased, not at a single point but from my shoulder blades to the small of my back. Pushing, pushing. I sank into my mattress and some air was expelled from my lungs. At that point I nearly did panic with fear. I hadn't thought that it was a person, but now the weight on me had me wondering if someone hadn't broken in through one of the downstairs windows.

The next thing I felt was inside my body. A sadness that coursed through me as if intravenously. That's the only way I can describe it. My body, my heart, were utterly infused with melancholy. An unclaimed grief washing through my chest, my limbs, my throat. Then it was pouring out of my eyes. As I lay pinned under whatever held me, tears gushed forth. I did not sob or cry out, but within seconds my pillow was drenched. Then my body trembled. And it was the strangest thing, because that seemingly bottomless well of despondency that had overtaken me was not my own. It was another's sadness roaming my bones. Someone else's tears.

There is no real way to explain crying someone else's tears. The hollow, unanchored agony. Then it stopped. The pressure on my back dissipated, as did the aching and the fear. I lay there against the dampness of my pillow, breathing heavily with relief, and could think of no compelling reason to sit up. I nearly laughed with the happiness born simply of no longer being sad. As I drifted back to sleep I forced myself to whisper once, "Come back."

But whatever it was, it did not.

When I awoke again it was morning. I dressed and went to catch my plane to Colombia. In just a few days' time the United States would take its military campaign against cocaine to an unprecedented level. Between Plan Colombia, the high-level talks with Mexico on immigration reform, and a renewed White House–led push for a free-trade zone spanning all of Latin America, George Bush was steering the United States—for better or worse—into its deepest engagement with the region since the cold war. And here I was poised to cover it.

Several hours later my plane landed. It was midafternoon and cloudy and cool in that highland capital. As I was checking into my hotel I thought, President George Herbert Walker Bush had the Gulf War; his son George Jr. has found his fight here, in Colombia. His Century of the Americas was truly getting underway, and I couldn't think of a single thing that might derail that. For dinner that night I had seared swordfish with a pleasing anchovy and garlic sauce. The date was September 10, 2001.

PART TWO

I was passing through the fifth security checkpoint in Bogotá's El Dorado International Airport. Pat-down after pat-down. If someone rubs my pant leg one more time, I thought, I'm going to develop a friction rash. A policeman was thumbing through my passport, looking for the page with the photo.

"You're American," he said.

"That's right."

"I'm very sorry."

"Thank you."

It will be pathetic if I start crying here, I thought.

In the line of passengers parallel to mine a young man in a pink button-down shirt took off his book bag and emptied his pockets, also for the fifth time. He was clean-shaven and handsome and privileged-looking. He looked like he could have been on his way to his first day of college in the United States. He and I passed in unison through what appeared to be the airport's final metal detectors. Neither machine beeped. I picked up my satellite phone case and audio gear bag, he his books, and we approached the waiting area together. Finally we'd be allowed to sit down. But before we'd taken our seats a policeman approached us from behind and tapped the man on the shoulder.

The young man turned and looked at me. I gestured with my eyes. He realized then that it had been the policeman and he bolted. He ran straight back through the metal detector, gunning for a closed emergency exit door. His outstretched arms were about to reach the release bar when he was tackled by no less than three policemen. The scrum of them fell forward, smashing through the door, setting off a loud alarm. Several other police officers crowded through the doorway, and for a brief moment I could see them swarming over the young man as if he were on fire. Then one of the police shut the door. The alarm stopped.

I found myself just standing there then, by myself, watching for something else. For someone else. I almost felt as if I should empty my pockets for the police again. But the officials who had not pursued the boy were not at all interested in me. They were already back at work, frisking more departing passengers. In a matter of seconds everything had apparently gone back to normal. This in itself was nearly as unexpected as the college kid's desperate ditch. Well, I supposed, it was just another day at one of the busiest cocaine transit points in the world.

But it was more. The incident was also an up-close reminder of what should have been obvious. Though the flames were still keening up through the ash-gray clouds of soot in lower Manhattan, there was still product to move, noses to feed, money to be made. There was no reason to think that cocaine smuggling should stop just because the United States had been attacked with airplanes.

Seven days earlier, on September 11, I was lying in my bed, in my hotel. Secretary Powell wasn't due in Bogotá until the afternoon, and rare was the opportunity to loll under the covers. It was nearly nine in the morning. Then the hotel room phone rang. It was my new editor, Didi Schanche, calling. She'd been hired to replace the departing Paul Glickman, the editor who'd ordered me to Haiti my first week on the job and with whom I'd been working since.

"Sí?"

"Gerry, it's Didi."

"Oh, hi."

"Are you watching CNN?"

"I hate CNN."

"Turn on the television."

"This better be good," I said.

The following morning I was in a cab driving through a numbing drizzle toward an exclusive neighborhood of Bogotá, higher up in the mountains. I was going to interview a political scientist at a local university for a reaction piece from Latin America. I was grateful for the assignment. Secretary Powell had of course immediately canceled his visit to Colombia in order to rush home to watch the Pentagon burn. Despite my own desire to rush home to Mexico, my editors had asked me to sit tight for the time being. I thought "for the time being" would mean a day or two.

As I directed the cab toward my rendezvous with the professor, the driver politely interrupted me.

"Excuse me, where are you from?" He was smiling expectantly in his rearview mirror. "Your Spanish is very good."

"It depends on how tired I am, but thank you," I said.

"But where are you from? You are not Spanish."

"No. I am from New York."

We were stopped at a red light and when it turned green the driver pulled his car over to the side of the road. He did not flip on his warning lights but just parked diagonally across the rightmost lane. Cars behind us honked. He ignored them. He threw an arm over his bench seat and swiveled around to look at me.

"I am very sorry for what has happened," he said.

"Thank you. So am I."

"We know what war is like," he said. "We know how it is to lose our own children. We've been fighting among ourselves for over forty years." Improbably, then, he reached out and took my hand. He literally had to fish it off my reluctant lap. This big, gruff-looking guy. He had one of those wooden-bead slipcovers over his seat to make it more comfortable. I did cry then. Like many New Yorkers, I knew people who worked in and around the Twin Towers. And we did not know yet who had died or how exactly. Burning or falling. But it was the cab driver's gesture that took my knees out. As I tried to control myself he just held on to my hand without saying anything. Cars behind us continued to honk. Finally I pulled free and stammered—in decidedly less than perfect Spanish now— that if we didn't get a move on I'd be late for my interview.

When I got back to my hotel room I found an envelope slipped under my door. It was a letter on hotel stationery, from the hotel manager, directing his sympathies to me. I thought I might go find this man. To tell him something. But I didn't know what. I felt a little bit lightheaded, so I decided instead to go for a walk, maybe interview some more people. On the first corner there was a man selling candy and cigarettes.

"*Me llevo una chupa-chup roja,*" I said. *I'll take a red lol-lipop.*

"*¿Cómo?*"

"One of those," I said, pointing.

"Here you go, take it," he said. Then, to my dismay, "Where are you from?"

"Scotland," I said.

"I think it's tragic," he offered anyway. "No country in the world deserves what has happened."

"No," I agreed, and moved on.

Over the next week I was slowly overcome by melancholy, this time my own. I found it difficult to concentrate and much easier to sleep. My state of mind, however, did not at all affect my work because I didn't have any. My editors still had not decided what to do with me, which meant that I was stuck here. In limbo in the Andes.

For the next seven days I wandered Bogotá's streets, a little bit lost myself. I rolled tape for the sake of it, interviewing people as a means of not losing sight of my job and as a way to conjoin my state of shock with a larger world in shock. Then Schanche called to say I could fly home to Mexico. I had produced exactly one story for all of those hours. And even that one had yet to air. I checked out of my hotel, took a cab to the airport, and got in line for my first security frisking.

In Mexico, until further orders came in, my plan was to pick up where I'd left off in Colombia. Microphone in hand. Radio therapy. My spirit was heavy and more than anything I craved the sort of collective mourning I'd found in Colombia as a means to shake this. That was the idea. But when I got home I found that many Mexicans were celebrating.

The cheers in the local cantina were deafening. We were watching the videotapes of the two passenger planes slamming into the World Trade Center towers over and over on the news. The Mexican newscasters wore expressions of disgust and awe. But for the Mexicans in the cantina each time the fireballs erupted it was like fireworks.

"*¡Toma, gringos de mierda!*" *Take that, American pieces of shit!*

"*¡Qué se chinguen!*" *Go get fucked!*

"*¡Ahora sí les toca, cabrónes!*" *Now it's your turn, you bastards!*

Amid the hooting and whistling and table thumping, Michel was staring down at his beer, embarrassed either for me or for the Mexicans or both. Michel was a friend of mine, a Chilean sound engineer and musician whom I'd met early on in my stint. He lived just down the street, and I'd called him right away when my plane arrived from Bogotá.

"*Pinche* Gerry," Michel said. "Welcome back, *cabrón.*"

"See you in twenty minutes at the cantina, *cabrón?*"

Now Michel was suggesting we leave. He offered to take me to a big party somewhere in the neighborhood. He assured me there would be lots of women and dancing and we could immediately commence a night of remorseless carousing.

Inside the party I saw that Michel had been right on most fronts. There were lots of pretty women, Mexican and otherwise, and everyone was dancing. When we walked in, a group of about twenty people was doing a line dance popular at the time that involved imitating a deer. Index fingers pressed to the temples like antlers. Shuffle to the left, shuffle to the right. Michel shoved a beer into my hand.

"*¿Perdiste a alguien?*" he asked.

"I must be the only New Yorker who hasn't lost someone," I said. "At least as far as I can tell. They're still searching for bodies."

I inhaled deeply. Here we go, I thought. But as I was sluicing back that first beer I realized that no matter how much I might drink I would still be the only one drinking over this. The rest of the glasses in that big living room were raised high to life, not to death. That was depressing. I put my beer on a table. I tried to dance once with a girl who smiled at me. Then I went home.

For the next several days the Mexican newspapers were filled with op-eds and commentaries on how justice, albeit tragic justice, had been served. On how naïve the Americans were for not seeing it coming. On how sad it was that people had died, but what did the Big Bully Up North expect after pushing the entire planet around since time immemorial? On how difficult it was to empathize with the wounded giant. On conspiracy theories as to who was really behind the attacks.

One night during that difficult period I was having a drink with a group of friends in a bar. The topic of New York came up in conversation—not, this time, related to September 11 but as a place to shop. Being still hung up on the attacks, however, I steered the conversation back.

"A lot of stores must be closed downtown these days," I said.

An Argentine woman among us lit up.

"*¡No me sorprendería!*" she said. "*¿Fue bárbaro, no, eso?*" *I wouldn't be surprised! It was incredible, wasn't it?*

"What was?"

"When those planes hit . . ." She was smiling with nothing short of authentic pleasure as she raced her hand across our faces in imitation of either American Airlines Flight 11 or United Airlines Flight 175. But nobody else was smiling, which is likely what caused her to remember that there was a *gringo* present.

"I'm sorry," she said, embarrassed.

The next day some Argentine friends staying at my house tried to explain where their compatriot was coming from.

"The United States has been screwing over Latin America for centuries," Guadalupe told me. Guadalupe was a fashionable young woman, in town to organize a musical festival. Her parents had fled the U.S.-backed military dictatorship in Argentina in the early 1970s. They'd been hippies and lefties, and had they not left, they might well have been tossed from a helicopter into the ocean. Guadalupe had grown up in Vancouver, Canada. We'd met during my years in Seattle.

"The people in Colombia were very kind," I said.

"The Americans didn't do in Colombia what they did in Argentina," she said.

"But the people in the Twin Towers weren't behind all that," I said.

"The American government was behind it," said Walter,

Guadalupe's husband, "and the American government responds to Wall Street. It's all interconnected." Walter was from Mendoza and was quick with a joke but now he was serious. "Many more Argentines died during our seven years of U.S.-supported military juntas than Americans on September 11. There is a lot of anger still. A lot of unhealed wounds."

That evening I wandered into my small shrine-room off the guest bedroom. I filled the seven prayer cups with water and lit the candles. I lit the incense stick in the candle flame. Then I sat and raised my two hands in prayer and said out loud the prayer about wishing to benefit all sentient beings. But my restless mind and my restless body proved too much. Five minutes later I was blowing the candles out again. As I backed out of the room I bowed to a photo of my teacher, Dzongsar Khyentse, hanging on the wall. "I'm sorry, Rinpoche," I said. "I'm sorry, but I'm too confused to be of much benefit to anyone now. Even to myself."

I went down to the kitchen and poured out the last swig of mescal from the unlabeled bottle I'd found when I'd first moved into the house. I poured it into a baked-clay coffee mug and inhaled its smoky fumes. Then I knocked it back. The process of distilling mescal begins by ripping out the heart of the maguey plant. That such deliberate savagery could age in time into something so exquisite struck me as somehow hopeful. It made me think of the meditation I'd just skipped out on. Because when you relax your mind your most savage memories are often the first to accost you. And then, with time, they soften. This might have been my cue to head on back upstairs to my meditation mat, to sit out this storm with patience. But I just couldn't.

I sat on my roof under the vapory sun, thumbing despondently through the pile of newspapers that had arrived while I was in Colombia. On the op-ed pages of the September 18 edition of the centrist paper *La Reforma* I came across an article that caught my attention. It had been written by a well-known Mexican journalist and commentator, a middle-aged icon of Mexican society named Guadalupe Loaeza.

In her commentary, Loaeza cast shame on her fellow Mexicans for not joining the collective voices around the world offering solidarity to the Americans. The French, the Chinese, the Russians, and the Canadians had all put aside their differences with the United States to extend a hand of friendship and an offer of help, humanitarian and otherwise. Loaeza accused Mexicans of cowardice, of hating without thinking, of being hypocrites. This was the first such article I'd seen.

How often do you go to New York shopping? she asked specifically of Mexico's moneyed intelligentsia and opinion makers. How many of you have second homes in Texas, in Florida? Who among you doesn't rush out to see the latest Woody Allen film? And enjoy a Coca-Cola while you're at it? Who doesn't jump at the chance to teach at a U.S. college?

And have you forgotten how the Americans helped us after the 1985 earthquake that collapsed much of Mexico City?

But her arguments, she wrote, found little purchase in the hard-baked soil of Mexican resentment toward the United States.

Haven't you read Noam Chomsky? her friends and colleagues allegedly chastised her.

Those towers were the very symbol of American arrogance.

The Jews were behind it.

The American far right was behind it.

Have you forgotten three hundred years of *gringo* disdain toward Mexico?

Have you forgotten how the Americans mistreat Mexican immigrants every day?

Loaeza ended the essay by saying that no matter how she reasoned she ended up getting too emotional and therefore lost her power to persuade. The attacks on the United States had touched something deep within the contradictory nerve center of admiration and rancor felt by so many Mexicans. So, she wrote, she'd decided to act instead. She said she was going to buy a wreath of flowers and lay it in memorial on the steps of the U.S. embassy downtown.

I pawed through the next day's *La Reforma*, then the next's. There she was again, on September 20. I began reading. Loaeza had indeed gone to the embassy, with a dozen colleagues, dressed all in black and carrying white flowers. And when they arrived they found that they weren't alone in their grief. Piled before the embassy were "clouds of flowers, as if sprung from the depths of the earth." Burning candles, bouquets. The largest was laid by the "Heroic Corps of Mexican Firemen." There were dolls and paper flowers and messages of condolence written in chalk on the sidewalk. There were images of the Virgin of Guadalupe, the religious mother and protector of all Mexicans. And there was a big sign from the Americans: THE U.S. EMBASSY THANKS YOU ALL FOR THESE OFFERINGS OF CONDOLENCE.

I walked to the embassy then. But more than a week had gone by, and all I could find was candle wax pressed into the pavement by passing shoes. I tracked down Loaeza's telephone number and called her to see if I could interview her. I went to her enormous house in the Polanco neighborhood and was met by a servant, who led me into a small parlor where tea was waiting but Ms. Loaeza was not. I took a seat. Finally she appeared, impeccably dressed and painted up like the women of her social class.

"The reaction here to the attacks has surprised me," I said. "I was expecting something else."

"What were you expecting?"

"Solidarity. Not its opposite."

Loaeza sipped her tea. "It is because we hate you and admire you," she said kindly. "We behave like the servants in a large estate of which you are the owner and master. We cannot shake this servile attitude. We do not like it, but we don't know how to act otherwise. We have a terrible inferiority complex. And we blame you for it. But you are only partly to blame.

"So what happens? Here is an example. You, the master, are heading to your room with a tea service and you trip and fall, spilling everything on the floor. Before you, for your benefit, we servants will rush to your assistance. But when you are gone, behind your back, we will giggle like tickled schoolchildren. Since September 11 we've been giggling. And this is what shames me most."

For her stance Loaeza was taking a beating in Mexico's press, especially in the dominant, center-left newspapers. I had to respect her for sticking her neck out. And I had to thank her. Because, quite simply, I left her house feeling a million times better.

I went home feeling more ready to work than at any time since the attacks. Like the rest of the world, Latin America was abuzz with speculation about how history might now unfold. Would the United States be hit again? Or would one of its allies—even one in Latin America, for that matter—be next? As the United States geared up for war, the discussion down south grew louder and more boisterous. How ought the Spanish-speaking world contribute to such an effort? Was war even justified? Perhaps, as many argued, Latin Americans should just lie low and stay out of a fight that most felt wasn't theirs to begin with.

I began pitching story after story. Hang on, hang on, came the message from my editors in Washington. The dust needs to settle. Finally they got back to me. They wanted reports. But soon a pattern developed that didn't bode well: the stories they commissioned all had to do with Latin America being put on the back burner.

I was leaning against a stretch of the corrugated metal barrier that formed the U.S.-Mexico border south of Laredo, Texas. Next to me was a man in a light green cowboy hat. He chewed a piece of grass of the same color. We were watching the sky for helicopters.

"What's your name anyway?" I asked.

"Why should I tell you my name?"

"It's a custom among journalists to ask for the names of the people we talk to."

"Our police have the same custom," the man said. He banged his elbow lightly against the metal border. "So do the Americans. Among us immigrants the custom is to keep as quiet as possible."

It was a quiet day in general. But on the American side of the border a massive security buildup was underway. Thousands more border patrol agents were en route, backed by all-terrain vehicles, mobile robotic watchtowers, and surveillance helicopters. A month had passed since the September 11 attacks and the Bush administration was scrambling to secure the perimeter.

"Will you try to cross again?" I asked the man.

"*Yo creo que no*," he said forlornly. "Getting caught twice in three days is enough." He picked up a canvas satchel and slung it across his chest. "They say the *gringos* are getting tough because of the terrorist attacks up there. Crossing right now is impossible. A friend of mine and I are going to try to scrape together the money to go home. I've lost hope. I just want to return to my town and my family."

"Where are you from?"

"There, yonder," he said, waving an arm. "Three days' journey at least."

"And once home, what will you do?"

"Fret as I did before, I suppose. Maybe work some if there's work."

"Aren't there less-patrolled places to cross?"

"Not for me there aren't. Not anymore. I am forty years old. I'd never make it."

The man was referring to the long, unwatched stretches of border, far from the major frontier cities like Nuevo Laredo, stretches that would likely never be thoroughly patrolled. The border was just too long. So the harshest areas, the most unforgiving terrains, were simply ignored by American authorities. The baked and desolate mesas, arroyos, and ridges where ground temperatures at midday could easily reach 140 degrees. Where the corpses of immigrants who could not handle the three- or four-day hike to a safe house were later found charred black by the sun's heat alone.

"Do you have a pen?" the Mexican man said to me.

"Yes."

I handed it to him and he crouched down on his heels, scanning the metal barrier like a man searching for secret runes. Finally he stopped at one spot and wiped it clean with the heel of his hand. Then he began to inscribe something. I thought, so I will have his name for my story after all. But the ballpoint pen wouldn't write on the metal. The man stood.

"*Lastima*," he said. "What a shame. I wanted to leave proof."

"Of what?"

"That I made it this far."

"What do you have to prove?"

"It's for others from my village to see. For my kids to see one day. For my wife. She is going to be angry when she sees me."

"I doubt that."

"Maybe not at first," the man said, handing me back my pen, "but when there's no food or money and our kids can't go to school, she will become angry."

"It's not your fault," I said.

"If you hire someone to build you a house," the man said, "you do not pay him to tell you all about his hardships. You pay him for the house."

"Yes."

"Marriage is the same."

"It's not a contract in the same way."

"If you say it isn't, then you have never been hungry."

"Not in the way you mean," I said. Then I had an idea. "May I go with you?"

"I just told you I'm not going."

"Not to the U.S. Back to your village."

"You would only bring me trouble."

"You could tell me your story."

"I've already told you too much," the man said. "*Vaya con Dios, Señor Periodista.*" Go with God, Mr. Journalist.

As I stared out over the bay of Port-au-Prince it occurred to me that Mexicans were now stuck in the same boat as Haitians. That is, their odds of escaping from poverty had just been reduced dramatically. The Mexicans were sealed in at home by a newly toughened northern border. The Haitians had long been imprisoned by the sea—the sea I was looking at now, in February 2002.

I turned my back to it and approached the enormous, two-story cinderblock building that stood alone on these flats on the outskirts of the capital. WELCOME TO DEMOCRACY VILLAGE, read a sign above the entrance. The drunk who'd been shadowing me since my arrival here followed after me still.

"Hey, *blanc! Blanc!*" he said, nearly spitting in my face. *Blanc* was the term many Haitians used for whites. "Hey, *blanc!* What are you gonna write about us? Do not shame us! We are just trying to live! Why don't you give me some money instead of coming here to make money off of us?" The man's eyes were wet and yellow. Absurdly, he carried his bottle in a paper bag as if there might be police around to enforce some drinking-in-public ordinance. But there were no police here at this poor squatter's hovel. No police or public officials of any sort. There never had been.

"I'm not going to make money off of you," I said. "I'm here to do a story on how your president is faring."

"What president?" the man slurred. "You mean the man in the National Palace? He stopped being my president a long time ago. He's not my president until he keeps his word."

"And what is that? To give you jobs? Schools? Roads?"

"No," the man said, steadying himself with a hand on my shoulder. "He promised that he'd come back one day."

We were now at the entrance to the building. There was just one way in, a shadowed chasm through which a truck full of prisoners might easily pass. Inside it was dark and warrenlike. A constant traffic of people in and out of the shadows. We walked past cinderblock walls lined with rows of cement cages barely three feet high. They were occupied now by children playing hide and seek. But for over three decades they'd held the hunched, breaking frames of untold thousands of political dissidents. This "village" was once the Duvalier dictators' most notorious prison. François "Papa Doc" Duvalier and later his son, Jean-Claude, aka "Baby Doc," ruled Haiti as strongmen from 1957 to 1986, with support from the United States.

My drunken guide sat down in one of the cages and pulled his knees up against his chest.

"This is how they fit," he said, swigging on his bottle.

The building was just a shell now. Even during the day its high windows, without glass, let in just enough light to mimic the dusk. Smaller rooms and corners of the larger ones were cordoned off with sheets to afford a bit of privacy for families. There must have been two hundred people living inside. There were no jailers to keep them here. They were trapped by circumstance.

I walked back outside. On the cement patio in front women were preparing cakes of clay in the sun, seated on their rumps, their legs splayed forward and wide as if to deal out cards among them.

"Sometimes we have salt, bouillon, or butter to add," one woman told me.

"These are for eating?" I asked.

"Whoever is hungry, he or she will eat them."

Then a group of small children was pulling at my arm, laughing and urging me out toward the sand flats. The whitened flats tufted with pale green beach grass sat between the promising sea and this dismal dwelling.

The children were laughing and skipping, and when we reached an area of open sand they began walking about in a crouch, filling their hands with shards of white bone. The tiniest looked so fragile, splintered like the tapered shafts of gull feathers, but the bigger ones were unmistakable. Fingers, forearms, and knees and, across one boy's arms, a bleached femur held like driftwood. Bones of all sizes jutted from the sand. I wondered how deep they went, how far you would have to dig to reach a time before the Duvaliers and their incessant executions.

The boy with the leg bone was trying to get my attention. "Sometimes the zombies, the people who died here, they get up and they walk around," he said excitedly. Then, like an invitation, "You could come back here to see them if you want." This small kid, oblivious to the tragedy enveloping him, holding a human leg and inviting me to watch the ambulating dead. The insult of it all, and the heartbreak.

"Thank you," I said, at a loss. "I would like that. I will do that." I couldn't help but feel like President Aristide himself with my promise.

We walked back to Democracy Village. Some people still called this place by its old nickname: Inferno of Man. A woman like a skeleton in a red dress took up beside me, staring at my recorder.

"How did you come to be here?" I asked.

"Because of the violence," she said.

"Where?"

"In Port-au-Prince. After a while you only want peace. Many others here tried to find it by boat. But they failed. And they washed back ashore here."

"And here it is peaceful," I said.

"No! It's not peaceful!" yelled the drunken man. "It's very far from peaceful! When you are hungry there is no peace."

"But no one is killing," the woman said scornfully. "Be gone, you!" She waved an arm at my self-assigned guide and then she laughed at him.

"Don't let him bother you," she said.

"I understand your president has a promise to keep."

"I don't know about that," she said. "I've only been here a few years. They say he came when he was first president. He changed the name to Democracy Village. I know that. They say he promised that the changes in Haiti would begin here. But do you really care what politicians say? Do you really waste your time on that? What are you anyway?"

"He's a journalist!" the drunk said. "Don't you know anything?"

The two began to argue loudly. I asked Karla what they were saying.

"The woman doesn't understand why all of this would interest you," she said.

"Tell her that sometimes my job is to follow up on promises our leaders make. As pointless as that might sound."

As the discussion in Kreyol continued I looked back at Democracy Village. That it was still standing spoke directly to the cynicism and laziness of Haiti's modern political class, the ranks of which were swollen with characters the Duvaliers would have chopped in two were they still in power.

President Aristide, on his first full day as president, on February 8, 1991, had led a tree-planting ceremony here in the company of French first lady Danielle Mitterrand. And he'd promised that he would convert the place into a new symbol. A symbol to lift the spirit, not to crush it.

Instead it had become emblematic of human inertia, disregard, and hypocrisy; by the time Aristide returned to the National Palace in 2001 he himself had grown used to living in a most splendid mansion, set in a decidedly more upscale Port-au-Prince neighborhood. And if he'd penciled this project in somewhere, if he'd remembered his promise to return one day, to transform this place into something better, it seemed he'd now forgotten.

He did have other things on his mind. His first year back in the palace was coming to a close and there had already been two apparent coup attempts against him. That had set off a wave of violence against independent and opposition journalists across the country. Aristide came under criticism for not doing more to protect them, and free speech. Soon, the Paris-based organization Reporters Without Borders came to Port-au-Prince to denounce the rapidly deteriorating situation. They were chased out of the country by an angry pro-Aristide mob.

After leaving Democracy Village I went to talk about the situation with one of the country's most respected journalists, Michelle Montas. Montas worked at Radio Inter and was the crusading widow of slain fellow reporter Jean Dominique. Dominique had been a hugely popular radio personality who for decades spoke out against successive Haitian dictatorships and juntas. Somehow he had managed to stay alive. Until April 3, 2000. That morning he was gunned down by unknown assailants here at the entrance to Radio Inter. To date no one had been convicted of Dominique's murder. After Michelle buried her husband she returned to work at the station, scooping up some of the shell casings still littering the station parking lot like cigarette butts. She added them to a crystal bowl on her desk filled with deformed bullets and shells from various attacks dating back more than a decade. It was a reminder, she said, of the impunity against which she too had spent her life fighting. For the sake of my microphone she raked her long, slender fingers through them. They clinked like ice against each other and against her wedding ring.

"We have a grenade too, if you'd like to see it," she suggested. She said it as if she were offering coffee.

"Is President Aristide orchestrating a campaign of violence and intimidation against journalists?" I asked.

"No," she said. "There is no deliberate attempt by the government to stop expression. However, there are some groups close to the party in power that seem to be able to carry out threats against journalists. And I think this itself is sending a strong message that reporters in Haiti are in danger.

"But I think as journalists we should stop seeing ourselves as the belly button of the world. It's not just journalists under attack, but people who have been in the democratic movement all along. In Haiti we have a problem with freedom of association. That goes for unions and peasant groups too. The problem exists throughout the country. And a number of elected officials don't seem to know the limits of their mandate and how far they can go."

"Are you referring to the president?"

"I think we're seeing a weak president, an Aristide who cannot control a number of factions within his party. These factions are mostly responsible for the violence. As to whether Aristide wants to control them? He says he does. He says he wants to negotiate with the opposition. But at the same time the opposition's headquarters is being burned down. Does this mean that Aristide cannot control the groups in the streets? To me the answer is obvious: he cannot."

Nor could Aristide control the country's fiscal slide, because the matter was largely out of his hands. The United States and international lending institutions were still withholding hundreds of millions of dollars in aid and loans. Aristide had once said that without international aid he would fail. It was difficult to see how he might avoid his own prediction now.

In this ever tenser atmosphere ordinary Haitians were taking it on the chin. And in the belly. In an open meat market in Port-au-Prince a fifty-two-year-old mother of three named Matwal Tayeese paused between skinning goats to complain to me about the economy.

"The prices of everything from food to school are going up and up," she said. Madame Tayeese stood, in open sandals, in the detritus of animals. "I blame the economic crisis on the politicians. They're fighting among themselves," she said, "but God will have the final word. You can't trust anyone. The politicians are all just human beings.

"Only God can bring change," she said, tapping her chest with the broadside of her cleaver, momentarily depriving a droning cloud of flies of their sticky feeding post. "I'm talking about the God within you."

Good people like Madame Tayeese might very well turn inward, toward God, after losing faith in politicians. But what might happen, I wondered as I left the market, if others, many others, expressed their dashed hopes outwardly? Part of the problem was that the American government was not spending time on this question. It had decided to leave Aristide's hands tied, but it had not analyzed the potential consequences. It was too busy now planning for democracy in Afghanistan and Iraq.

By the spring of 2002 the cost of American inattention in the Western Hemisphere had become the leitmotif of what stories I could get on the air. Which was why I was sitting on the floor in a wooden hut with a dying baby named Teresa.

Teresa was a two-month-old Maya girl born in the arid mountains along the border between Guatemala and Honduras. She was dying of malnutrition in the arms of her mother. Her wheezing sounded like a handsaw cutting pine board. Lazaro Roque, my Guatemalan fixer, was with me, helping me interview Teresa's mother in her family's ten-foot-by-ten-foot hut, dirt-floored except for the rugs in one corner for sleeping and in the center where embers glowed within a ring of stones. The Indian woman was still not recovered from giving birth to her child and now she was watching her slip away.

"Where is the doctor?"

"In the health clinic," she told us.

"And why don't you go to the clinic?"

"Because the nearest one is a day's walk away, far below in a regional town. Too far away."

Teresa's mother was visibly aggrieved, desperate, but resolute in her position. Her refusal to budge was baffling at first, given the urgency of her baby's condition. But her situation was common, she told us.

"I do not have the money for such a trip. And I lack permission from my husband to make such a dangerous journey alone. These mountains are filled with thieves who would not hesitate to take advantage of a lone woman. Even a woman carrying a sick baby."

This was a part of the world that, months earlier, had stood to benefit from U.S. aid in the form of mountain clinics and traveling physicians who could have easily attended to Teresa and so many children like her. But such programs had been suspended in the aftermath of 9/11.

Lazaro was practically conducting the interview alone and I did not interrupt. This Indian woman was more wary of me than of her Guatemalan neighbor, not Indian himself but *mestizo* at least. Lazaro sat on the floor with his legs crossed and listened as the woman spoke. Her story was pointlessly tragic and could easily have ended somewhere else besides next to Teresa's grave. But that is where it would end. Because Latin America was on its own. Even though its U.S. embassies were open and operating, issuing statements on regional occurrences and pronouncing on events, without the president's attention and cold cash to back them up, their shadows did not reach far. And worse, it seemed that the money the U.S. government was still pumping into the region was being monitored less carefully. A big pile of that misspent money is what had me back in Guatemala now.

Over five million dollars' worth of cocaine had just disappeared from a police warehouse in Guatemala City. The elite officers from the Department of Anti-Narcotics Operations (known by its Spanish acronym as the DOAN) who were under suspicion of theft blamed the drug's disappearance on a gang of stray cats. They said the cats had snuck into the warehouse through a vent and had eaten the entire stash over the course of several weeks.

The story got wilder because the United States was funding this same unit now under investigation. And to make matters worse, as the scandal was breaking a contingent of those elite officers was busy terrorizing a village in eastern Guatemala, holding the townspeople hostage and ransacking their homes, supposedly searching for drugs. They stayed for a couple of days and shot dead two local men who tried to flee. The town was called Chocón. It sat in a swampy lowland hamlet on the edge of a vast river delta in the east of the country.

Lazaro and I were on the road again, in a truck I'd rented at my hotel. Chocón wasn't on my map, but I was now familiar with Lazaro's built-in GPS tracking system.

After several hours we arrived at a small town on the edge of the vast estuary that eventually leads to the coastal town of Livingston. We walked to the water's edge where several small locales rented boats and offered tours of the wild and expansive waterway. Lazaro asked one young man if he knew how to get to Chocón.

"Chocón?"

"The village where the DOAN shot up everybody."

"Yeah, sure," the guy said and called out to a colleague. Another man came running over. "Take them to Chocón," he said.

We forked over forty dollars to the first young man and climbed aboard a small motorboat. Three hours later, after speeding through a confusing maze of narrow, mangrove-lined waterways, we still had not found the village. It was nearly dark and Lazaro was fuming. The trip was supposed to take fifteen minutes.

"You said you knew where Chocón was!" he finally exploded.

"I never said I knew where it was," our captain responded nervously. "It was my friend. But I figured I could find it."

"You're lying," Lazaro said. "You just wanted our money." He made his way to our captain and grabbed him by his gray sweatshirt. "You're going in the water," he said.

"No, wait, *por favor*, I swear! There was a misunderstanding!"

When we returned to the docks the guy we'd given the money to was nowhere to be found.

"I don't know where he is," the captain said.

"Where does he live?" Lazaro asked.

"I don't know."

Lazaro slammed him up against our truck.

"Easy," I said, "Hertz will charge me for any damage."

"Okay! Okay! I'll show you his house."

"Let's go," Lazaro said. Then he turned to me. "Not you," he said. "You stay here. I'll handle this."

"Let's forget it," I said. "I'll write it off. You're never gonna find the guy."

"I'll get your money," he said. The long boat trip had given him ample time to nurse his anger, and it seemed there was no way he was going to let it go.

"I'm coming with you then," I said. "I'm your backup."

"Sorry," Lazaro said. "You wait in the truck. You don't know these people." He spoke with an air that made me feel as if I were working for him. I climbed in the truck and watched him march off, holding our captain by one arm.

An hour later I was nervous. I decided I had to do something. I started the truck and pulled away from the curb, thinking I might drive around and spy Lazaro on one of the small town's streets. But I'd gone less than thirty feet when Lazaro came running up from behind and pounded on the driver's window. He was grinning and holding the forty dollars.

He jumped into the truck. "Our captain escaped," he said, "but I found the hustler's house," he said.

"You did? How?" Lazaro ignored the question.

"I knocked on the door. His mother answered. She says her son isn't home. So I said, okay, I'll wait then. I turn around and stand facing the street. After fifteen minutes the guy's mother becomes uneasy. She must have thought, this strange man on her stoop is not going to leave. So she begins shouting upstairs. I look up and I see the son of a bitch on an upstairs balcony. He has a gun in his hand. *¿Qué, Mama?* he yells. Give us our money back, I yell up to him. Beat it, he says. He looked surprised to see me at his door! He waves his pistol at me. Then his mother is yelling, Money? What money? She can't actually see her son from down below so she walks out into the street. But the guy has already tucked his gun away.

"'It's nothing, Mama,' he yells. 'This doesn't concern you.' 'Are you in trouble again?!' she yells back. She's furious. 'You get down here and give this man his money back! Or I'll give you such a thrashing!' Finally the guy buckles under his mom's admonitions. He comes down the steps and hands her the money, she hands it to me, and here I am."

"Damn."

"We lost time," he said, "but not time *and* money."

We spent the night in a cheap hostel and found Chocón the next morning. It was a rural hamlet with a few wooden houses hidden in a swampy forest, a tin shack convenience store, and not much more. We began taking testimony from the traumatized inhabitants. As I was interviewing one weeping woman I noticed a bunch of flat, rectangular paper packages strewn along the forest floor. I picked one up. It was an empty MRE, a Meal Ready-to-Eat, a military food pack. I flipped it over and found a stamp that surprised me: United States Army. I slipped it into my bag.

The next morning I went to the U.S. embassy to interview the then U.S. ambassador, veteran diplomat Prudence
Bushnell. Lazaro had wanted to accompany me during the
interview, if not for anything else but to see the inside of the
embassy. But security wouldn't let him in. He sat outside in
the truck and waited, disappointed.

Bushnell had made her name at the State Department
in part by having survived the 1998 terrorist attack on the
U.S. embassy in Nairobi, Kenya, carried out by Al Qaeda.
She was hard-nosed and frank and her appointment to Guatemala by Bill Clinton in 1999 spoke to the region's impor-
tance in the pre-9/11 world.

She had on her desk a plaque that typified her approach
to work and diplomacy in Latin America. It read, in Spanish, LO QUE PASA ES QUE . . . The words were circled and
a red line ran through them. The translation goes, "What
happened is that . . ."

"When a Guatemalan starts a sentence with that phrase
you can be sure that he or she is about to give you an excuse
for not having done something," Bushnell told me. "So I've
banned the phrase from my office." I thought it was funny,
but I wondered what her local staff made of it.

I handed her the U.S. Army food packet.

"This came from Chocón," I said. "Were U.S. military
personnel with the DOAN when they went on their rampage?"

"They were not," she said, looking as surprised as I'd
been. "But," she said, "the U.S. government is funding the

DOAN's fight against cocaine smugglers. I'll look into the matter more closely."

Several weeks later the United States cut off funding for the program.

The DOAN was disbanded.

Lazaro and I were on a roll. The next morning we were up early crossing the country in the other direction. We caught up with a group of Guatemalan forensic scientists searching for mass graves from the civil war period. It was to be the first dig in months. The excavations had been suspended because of death threats that unknown people had faxed to the offices of the Guatemalan Forensic Anthropology Foundation, the organization carrying out the digs. In Guatemala's decades-long conflict some two hundred thousand people were killed, 90 percent of them Indians at the hands of government military and paramilitary forces, according to human rights groups and a government commission. The forensic scientists were trying to find and identify those victims as part of a national process of reconciliation and healing. But some in Guatemala considered such digging an open provocation.

Today the shovels were breaking ground outside the tiny mountain town of Chichupac in north-central Guatemala. When we reached the site some locals were already there, waiting. Two policemen had been dispatched to stand guard during the excavation. They stood some yards away in the

trees, dressed in black, sucking on cigarettes. The crew spent
the entire day digging trenches, trying to find the exact lo-
cation of a mass grave that local Indians assured them was
there. No one spoke. The only sounds were the dull slicing
of shovels, the patter of falling dirt, the constant buzz of insects.

Late in the afternoon, as thick clouds rolled in and thun-
der began booming above us, one group of diggers made a
find: a collective grave. First to appear were the remains of
Eusevia Grave Garcia, an eighteen-year-old girl who villagers said had been dragged off by soldiers on a night more
than twenty years earlier. A group of five Indian women
huddled on the edge of the dig gasped as the forensic scientists delicately dusted off a gray lump that turned out to be
part of the girl's lower jaw. Then they pulled out her shirt,
reduced now to rags of tattered cotton but still bright red.

"That's my daughter," one woman said. She began to cry.
"I told you she was here," she said to no one. "I told you. This
exact spot."

"*Señora*," I said quietly. "You've known all along that your
daughter was here?"

"*Sí, señor.*"

"Why did you wait twenty years? Why now?"

"Because we are less afraid. Now there is peace. Before
1995, if you looked, you could disappear too."

"That's it for today, people!" called a scientist, setting his
shovel aside as a heavy downpour began. "I'm not shoveling
mud!"

Lazaro and I said good-bye and made the long drive back to the capital. It was late when we arrived.

"Just take the truck home," I told him, when we reached my hotel. "And come pick me up in the morning, at nine a.m."

But Lazaro didn't show up at nine. Or at ten. At eleven o'clock my hotel phone rang. *"Lazaro, dónde estas?"* I asked. "I've been calling you all morning."

"Gerry, it's Susana. Lazaro's, well, ex-girlfriend. You better come to my office right now. Some men kidnapped Lazaro. They were going to kill him."

The hair on my head stood up.

"What exactly happened?" I asked calmly. "Where is Lazaro now?"

"He is here, with me. He is rolling on the floor."

I ran downstairs and jumped in a cab outside the hotel. When I ran into the nondescript two-story building that housed Susana's production studio I found Lazaro, still on the floor. His face was contorted, his cheeks wet with tears, wincing as if someone were pointing a gun at his face up close.

"They were going to kill me!" he began to cry when he saw me. Tears were flowing down his face and mucus ran from his nose. "They were going to kill me!" I sat down next to him and held his head against my chest. He was shaking uncontrollably. His jeans were ripped and I could see he had a bad bruise on one knee.

"Who was going to kill you, Lazaro?" I asked. But he couldn't calm down enough to talk.

"*Shhhhhh, calmate, tranquilo,*" I said, trying to soothe him.

Susana and her colleagues were standing around watching like pedestrians whose morning had been ruined because they had witnessed a car wreck.

"Could someone get a glass of water?" I said, a bit angrily. "Has anybody called an ambulance? We've got to get him to a hospital right now."

"I would not recommend it," Susana said coolly. "It's too dangerous."

"What do you recommend then?"

"A private clinic, if you wish," she said. "At the public hospitals they register you under your real name. If someone is still looking for Lazaro they would easily find him there."

Susana picked up the phone and called a taxi. We hustled Lazaro into the back, and I climbed in next to him. Susana stood at the curb. "Aren't you coming?" I said out the window.

"No," she said. "It is better that you go."

"You guys must have had one hell of a breakup," I said.

But there was no time to argue. I told the cab driver to go.

In a matter of minutes we were at the doors of a brick medical center with a long entrance ramp for wheelchairs. I ran inside. A nurse followed me out, and we put Lazaro in a wheelchair and hustled him inside.

"What is his name?" asked the intake nurse.

"José Perez," Lazaro said weakly. José Perez is the Spanish equivalent of John Doe.

"No problem," said the nurse.

I was told to wait in the reception area as Lazaro was wheeled away. Later a female doctor came out.

"José is not injured badly, but he is in shock," she said. "It is not uncommon in these types of cases. He has been extremely lucky. Follow me, please."

She led me to his room where I found him in bed, in a white hospital gown, with an IV in his arm. He was sedated and calm. His eyes were open, staring at the ceiling. I wondered what he was seeing.

"Lazaro, can you talk?" I whispered, when the doctor left. "What the hell happened?"

"This morning, I went to my office before going to get you," he said weakly. His eyes were glassy and half open. "Four men with military haircuts and pistols just appeared out of nowhere. They surrounded me on the street. They took the keys and forced me into the back of the truck. They punched me and forced my head down between my legs. One of them kept me pushed up against the truck door. I could barely breathe."

"What did they want?"

"I don't know,'" Lazaro whispered. "I don't know. They kept saying, 'Give it to us or you're dead.'"

"What was it?" I asked.

"I told them I didn't have anything! I tried to explain that it wasn't my truck. They didn't believe me, or they didn't care. Then one of the men told the driver to go to El Gallito. And I knew I was dead."

"What's that? The Little Rooster?"

"It's our killing grounds."

"What the . . . ?"

"It's where the police usually find the bodies of people who mess with the real powers in this country. No one who enters El Gallito at the point of a gun comes out alive. Then the truck stopped at an intersection. I found the door latch with one hand. I managed to dive out. I landed on this knee," he said, pointing vaguely toward his legs. "*Puchica*, I was up and running before the truck had even stopped.

"I didn't know where I was going," he whispered. "I ran into a bank, but when the security guard saw me he threw me out and locked the door! Without thinking I ran all the way back to the office, to Susana." He was starting to get panicky again. I told him to rest.

When he'd fallen asleep I went out to see if I could get a pizza delivered. I ate it in silence next to Lazaro's bed. A couple of hours later he suddenly awoke and sat bolt upright. His hospital gown wasn't tied right and it slipped off his bare shoulder onto the sheets.

"The materials!" he said, wide-eyed. "They said they wanted the materials!" He looked around the room as if the materials might be there.

"What? What materials?" I asked, startled.

"They said, 'Your materials.'"

"What, our tapes? From where? Chocón? The dig?"

"I don't know. I can't remember."

"Try."

"They wanted something of yours. But I don't know what."

"Do you have your passport with you?" I asked.

"It's in my apartment. They took my wallet with my Guatemalan identity card."

I went out to talk to the doctor.

"I think my friend and I should get out of Guatemala right away," I said.

"Impossible," she said. "Your friend is not well enough to travel. He needs to rest."

I called Susana. "What should I do?"

"I don't know. What you shouldn't do is go to Lazaro's apartment to look for his passport. Not even with a police escort."

"Can you go?"

"Absolutely not."

"Someone's got to help Lazaro get out," I said.

"There are human rights groups. I'll get you their contacts."

I went back to his room.

"Lazaro," I said, "I think we both should leave the country as soon as possible."

"Look at me," he said. "You go."

There was a knock on the door. A young man tried to enter. I ran over and blocked the door with my foot.

"What do you want?" I asked.

"I'm Lazaro's brother," he said solemnly. I removed my foot and opened the door. The man took up a position near Lazaro's bed. I had never met either of Lazaro's brothers. He was shorter and rounder, and like Susana he seemed oddly unmoved by Lazaro's circumstances.

"Are you sure?" I said to Lazaro. "I'd prefer us to leave together."

"*Sí*. I'm sure. I'll come later."

I looked at his brother. "*¿Le vas a cuidar?*" *You're going to watch after him?*

He answered with a faint nod.

"Okay, then," I said. "I'm going to go back to the hotel and see about flights out."

Lazaro smiled, for the first time that day.

"As soon as I'm home I'll call Susana to see if someone's been able to get your passport. Maybe the police can help."

Lazaro's brother smirked.

"As soon as we can get you out of the country," I said, "we'll get you out. I promise."

I flagged a cab and went back to the Holiday Inn and paid my bill directly.

"If anyone asks for me," I told the girl at the reception desk, "please say that I've already left the country."

"Of course, Mr. Hadden," she said. She asked no questions. Were fleeing foreigners such a common sight? I went to my room and called the airlines. The earliest flight would leave the next morning at dawn. I stuffed my gear and clothes into my suitcase and ran half a block down the street to another hotel.

"Room for one. One night," I said.

"Name?" asked the male receptionist.

"José Perez," I said, looking over my shoulder.

"*Lazaro, estas bien?*"

He finally called me. I'd been home two days.

"*Mas o menos,*" he said. *More or less.* He sounded shaken. "I went home to get my passport yesterday. Two men were waiting outside my house. They saw me and they started following me. I ducked into a big shopping center and lost them in the crowds."

"Where are you now?"

"In hiding." There was a pause. "Are you going to get me out of this?"

"Yes. You stay where you are and you stay calm."

"I really need to get out of here. Any moment they could find me. And they won't let me escape twice."

I'd spent every waking moment since returning home trying to secure Lazaro safe passage out of Guatemala. Phone calls to human rights groups, the UN, the State Department, whoever would listen. A week later, it was Ambassador Bushnell who got him out. According to the ambassador's press aide, Bushnell went to Guatemalan president Alfonzo Portillo personally and demanded that Lazaro be given safe passage from the country. It worked. Eight days after his abduction he was on a plane for Mexico City.

"Man, am I happy to see you!" I said when Lazaro came out of the baggage claims area. He was smiling and baggy-eyed, and his clothes were wrinkled as if he'd been sleeping in them. "And I am relieved," I added. "It's over, man."

"I can't believe I'm out. Please thank Ambassador Bushnell for me."

"You'll have the chance to do that yourself. By phone, of course. Come on, let's get you cleaned up."

I took Lazaro straight to the house on Jojutla Street for a shower and a hot meal. I'd cleaned out the old bedroom that had belonged to Concha and her sons and placed a mattress with bedding on the floor. I'd also dragged up an old chest of drawers and made an impromptu night table out of a stack of newspapers covered with an old Indian tapestry.

"Here's your new home," I said. "It's a humble start, but at least you're safe."

"*De lujo,*" Lazaro said, setting his small bag down. *Luxurious.*

That same night I gave Lazaro a set of clean clothes and a few hundred pesos and took him to a gathering at the house of an AP reporter in the tony Polanco neighborhood of the capital.

"My goal," I told him as we drove, "is to find you work right away, either as a production assistant or a cameraman, whatever we can find. What you need now is to keep busy. To start making a new life. Going back to Guatemala is going to be out of the question for a while." Lazaro stared at me blankly.

"Don't look at it that way," I said. "Think of it as a new beginning. You're not alone."

At the party I introduced Lazaro to those who'd never worked with him, telling our story. Most people listened in disbelief. Lazaro was still in shock and remained silent most of the night.

In the morning when he awoke I was already at my desk in the next room.

"I am going to ride the bus today."

"Great," I said. "Where to?"

"To the end," he said. "Then I'll jump a different one. I need to get oriented."

He left early that morning and did not return until after dark. Two days later he popped into my office with an announcement.

"I've got most of the city bus routes and landmarks committed to memory."

"Your GPS is now loaded and up to date," I said.

"It makes me feel better."

"I know."

Meanwhile I'd been in touch with Mexico's National Human Rights Commission. A young caseworker there named Xochitl was assigned to Lazaro's case. The three of us went to lunch together for their first meeting.

"My job is to document your story," she said to Lazaro as we sat in a small bistro near my house. "And to help you in whatever way I can to adapt to your new life in Mexico." Xochitl was as chipper as a cruise ship hostess. Lazaro took to her immediately and soon she was regularly visiting the house.

At first she was coming by to record and counsel Lazaro through what was obviously a painful transition and recovery. But soon I began to suspect that her visits were taking on a more personal nature. Sometimes she'd come by in her car and honk, and Lazaro would leap up and leave. Once I came home to find them cuddling on the couch.

"Did I just catch the two of you necking?"

This went on for about a month. Then one day as I entered the front door Xochitl stormed out, barely saying hello.

"What's wrong with her?" I asked Lazaro, upstairs.

"She didn't like what she saw in the mirror," he said.

"You're not that bad," I said.

"She saw people walking around the living room. They were wearing old-fashioned clothes."

"Where?"

"In the mirror."

"She saw that in the mirror?"

"Only in the mirror," said Lazaro. "There was no one here but the two of us. We were just sitting there on the couch. Then she sat up and began drawing people on her notepad. Staring at the big mirror. She started describing how several well-dressed men and women were milling about the living room, around us, in silence. I didn't see anything. After a few minutes of this she just grabbed her coat and left. As you were coming in she swore that she would never step foot in our house again, no matter how urgently I might need to see her."

"That seems a bit over the top," I said. "Don't you think? Or maybe I'm just growing used to this shit."

"What shit?"

"Let's move the mirror, just to be safe."

"Safe from what?" he asked.

"Don't worry," I said as we lowered the heavy glass to the floor, then slid it toward the dining room. "The ghosts like journalists."

"So there *are* ghosts? *Puchica*."

"They probably would have liked Xochitl too if she'd given them a chance. That's the key to the whole thing," I said. "Anyway, I hope she comes back. I don't want your link to Mexico's human rights office to be severed by a misunderstanding with phantoms."

But Xochitl did not appear again. With that support mechanism suddenly gone I felt an ever more urgent need to find Lazaro work. But I was starting to wonder whether he was up to it just yet. A month had gone by since he'd arrived and he still seemed unmotivated and adrift, hardly the go-getter that had impressed me back in Guatemala. I went online and read up on posttraumatic stress disorders. The literature seemed to describe Lazaro's behavior perfectly. And a month was not much time to get over what he'd been through.

But as time went on he only worsened. He took to staying out very late and sleeping until noon. He had found a therapist but never went to his appointments. He would emerge from his room in the morning bleary-eyed and nervous. Nights filled with nightmares. Finally a friend and fellow reporter, Marion Lloyd from the *Boston Globe*, found Lazaro a job as a cameraman's assistant for a TV soap opera.

"It's not journalism," I told him, "but it will provide some structure and money."

He was without direction, working his way through the painful process of coming to terms with his near death, his new life, and all that he'd left behind, including his latest girlfriend, Sara, a print reporter for a Guatemalan daily. Work would keep him busy. It would be a beginning, at least.

One day soon after, Lazaro came into my office with some surprising news.

"I found my sisters."

"You have sisters?"

"Two sisters, half sisters."

"Where? In the mirror downstairs?"

"Here in Mexico City," he said.

I had lent him a small digital camera, which he now turned on to show me a couple of pictures taken over the last few days. In one photo Lazaro was smiling in a kitchen with his arms around both of them. "They're twins," he said proudly. In another photo the three of them were on horseback in what looked like La Marquesa, the green hills that border Mexico City to the south.

"So this is what you've been up to recently. That's fantastic. What are they doing here?"

"They left Guatemala many years ago, before I was even born," Lazaro said. "But I had their names and was able to track them down. They received me warmly."

That's just what was missing, I thought. Lazaro's life is going to take a turn for the better. But I was wrong. A few days later I saw Marion Lloyd from the *Boston Globe*.

"Lazaro's been fired from the soap opera gig," she told me over a drink.

"What the hell?"

"All I know is that he'd been 'acting strangely' on the set."

When I got home that night Lazaro was watching TV in the living room. "Hey," I said. "What happened with the job? Marion tells me they let you go."

Lazaro just shook his head.

"Did something happen? You wanna talk about it?"

No, he did not.

Two months passed but it was as if he'd just arrived. Lazaro remained listless and jobless. I'd given him the green light to fix up his room as he pleased—painting it, picking out furniture (NPR had generously agreed to pay him a small monthly stipend until he could get on his feet), and so on. Instead his room, the only space he had to call his own, was quickly turning into something akin to a homeless encampment. Garbage and dirty clothes and scraps of bean-stained tortillas were beginning to pile up. I took to keeping the door closed because of the smell.

During those weeks of dealing with Lazaro's unraveling I had the ongoing task of reporting on the same process now in full swing in Mexico. In the spring of 2002 Mexico was taking one tough turn after another. Kidnappings and bank robberies were on the rise as fewer Mexicans were able to slip into the United States to work. On top of that, over a hundred maquiladoras, foreign assembly plants, had shut down so far that year, moving their operations to China or India. Desperate Mexicans were forced to take homegrown jobs under horrific conditions. In July the southern state of Oaxaca actually sued Baja California for exploiting its mainly Indian residents on plantations it described as akin to slavery.

Meanwhile, President Fox was desperately trying to show the outside world that Mexico was indeed changing, modernizing under his leadership. To prove it he was trying to build a new six-runway airport outside Mexico City. But his multimillion-dollar nod to modernity was sabotaged by that most nonmodern of Mexican institutions—the peasant farmer. Local farmers rose up against the project. They threw up roadblocks and took riot police hostage. Fox's new airport never got off the ground.

Inevitably, when I'd return home from reporting these events, I'd find Lazaro sprawled out on the downstairs couch staring at the television. He was becoming as much a fixture in the living room as the piano. On several occasions I turned the TV off and tried to talk to him about what he was going through. But he would just shake his head nervously. That nervous energy became contagious. Guadalupe and Walter, my Argentine friends staying in the guest bedroom, began to complain.

"Can I tell you something?" Guadalupe said one evening. "I'm afraid to be alone in the house with Lazaro."

"Why?"

"He has an aggressive demeanor that makes me uneasy."

"He's very tired and nervous," I said.

"The other night he came into my room and asked me if I had any hash he could smoke. I told him no, but he lingered anyway for a long time in my doorway without saying anything. Then we heard Walter come in and Lazaro ran back downstairs."

"He's doing his best," I told her. "We have to give him time. He's been through a lot."

"He's creepy," she said.

Lazaro's skittish behavior only increased over the next month. I found myself growing frustrated.

"Lazaro," I told him, "we have to talk about your future."

We sat down in the living room.

"I confess that I don't know what to do," I said. "You've now lost a couple more part-time jobs. And you seem, well, you seem less fit to stand on your own feet than when you first arrived."

Lazaro's hair had grown long and hung in frizzy curls in front of his eyes. He had a delirious look about him, as if he hadn't slept in a long time. But he seemed to be sleeping all the time. He sat on the couch silently, his head bent.

"You've got to do something," I said. "You can't just stay here on the couch. You need to fight back."

"*Sí, sí,*" he said. "*Sí, sí, sí.*"

Time passed. He deteriorated further. I tried giving him small deadlines. I told him to write up a résumé within the next couple of days. He didn't do it. I asked him to return a pair of my pants that I'd lent him weeks ago. He balked.

"Where are they?" I asked.

"At my sisters' house," he finally told me, "being washed." Then Guadalupe came to talk again.

"Gerry, nearly three hundred dollars has gone missing from a bag in my room."

"Well, perhaps . . ."

"I don't lose money," she said, cutting me off. She gestured toward the stairs leading to Lazaro's chaotic bedroom.

"No."

"Who else?" she said. "Think about it."

Might Lazaro have stolen the money? At any other time in our relationship I would have said impossible. This was the same guy who could have gotten himself shot getting my forty bucks back from the boat hustler near Chocón. Lazaro

Roque was as tough as they come and a man of honor. But I had to admit I was beginning to have my doubts now. Lazaro had stopped talking to everyone in the house and was keeping to the back staircase, moving between his room, the front door, the couch, and the kitchen.

On the eve of a trip to Haiti I presented Lazaro with a final deadline. "If you haven't cleaned up your room, written a résumé, and decided to start talking to me again when I return in a week, then I am going to ask you to move out."

I came home a week later to find Lazaro asleep on his filthy bed like a drunk in a flophouse. When he stumbled to his feet later that morning I asked for his résumé. Negative.

"Have you found a job?"

"No."

"Have you looked?"

"No."

"Where are my pants?"

"At my sisters' house."

"Could we talk seriously?"

Lazaro just stared at the ground.

And that's when my patience ran out. "You have twenty-four hours to wake up," I said sternly. "Just a little bit. Some first step. If not, I'm changing the locks on the front door. I can't live like this anymore. Do you understand?"

He turned his back on me and went into his room again.

"I don't know what else to do!" I called. "I don't think I'm helping you anymore by letting you stay here! That's the honest truth, Lazaro!"

It was midday. Lazaro got dressed, slipped down the back

stairwell, and left the house. I did not see him again until the following evening.

Some fifteen people were over for a weekly movie club I'd started. Each week somebody different would pick out a film and we'd gather on the couch and floor and watch. Outside it was pouring rain. As we were all intently staring at the TV screen, the front doorbell rang. I knew who it was. That afternoon I had kept my word and changed the locks on the front door. Lazaro's key was now useless. I stood up. This, I hoped, would be the confrontation that would loosen the tortured air of tension that had enveloped the house. This might just jolt Lazaro out of his funk.

I went to the front window and looked down toward the door. But no one was there. Then I looked down the street. Standing under a streetlight a hundred feet away, soaked to the bone, stood Lazaro alone. His legs were spread wide. His arms hung at his sides as if he'd lost the use of them, and his head was hanging against his chest. I ran downstairs and opened the door, suddenly consumed by guilt and worry.

"Lazaro!" I screamed. "Get your ass in the house!" He didn't move. I began to walk toward him, but when he sensed me nearing he retreated farther down the street. "Lazaro," I yelled, "you're soaked. Come inside and dry off. We can talk some more."

Finally he lifted his rain-soaked head. He raised his hands to his temples and shook his head back and forth like someone receiving terrible news over a headset. Then he looked up at the sky and screamed a long scream at the top of his lungs. A high-pitched wail that sounded young and girlish and full of terror.

I ran for the house.

"I need help," I announced to my guests. "It's Lazaro." Three friends came with me outside to try to talk Lazaro down. He was stamping up and down now, in a blind rage. He ran to the curb and began pounding on cars with his fists, kicking them with his waterlogged boots. He banged his head against the side of a house and fell down in a puddle, still wailing.

Lazaro, we were all saying as we moved toward him. *Lazaro. Calmate. Calm down.* He must have thought we were going to grab him because he ran farther down the street.

That's when a French woman named Anne appeared at my side. Anne was the young wife of a French ship broker who'd come to seek his fortune in Mexico. They'd been in town just a few months. In that time Anne had already picked up Spanish. She'd learned it purely informally, through her daily haggles with the cagey *señoras* ensconced in the stalls of the capital's myriad *tianguis*, or outdoor food markets. It was like learning English from Coney Island carnival ride operators. She had absorbed the accent of the streets.

Her long brown hair was already flattened against her head by the rain. I offered her my jacket.

"Let me try to talk to him," she said.

"Why not?" I said. "Just be careful." But I wasn't overly worried. Lazaro knew Anne from her visits to the house, and he seemed to like her. The men went back inside, and

Anne approached Lazaro alone. I watched her from the doorway, ready to sprint to her aid if necessary. That was one reason.

After just a few seconds of talking, Anne somehow got Lazaro to stop stomping about and to sit down. There they sat, together on the curb in the downpour, neither of them paying the slightest attention to the sheets of water unfurling against them—she speaking, he nodding his head solemnly amid the hiss and spray of the rain. Then he began to wave one hand anxiously, pointing back toward the house. With his other hand he wiped his hair over and over from his face. He appeared far from pulled together, but at least he wasn't running away or hurting himself further. I didn't know what Anne was saying to him, but I suspected the words were secondary. The main question was her presence. By virtue of just being there, of being who she was, Anne was cutting through Lazaro's mad, hermetic solitude where the rest of us had failed. She had reached him, even if just for a moment. I was quite sure of this because she had a way about her. Because each time she came to my house she had the same effect on me.

The other reason I was watching Anne talk Lazaro down was because I was horrendously in love with her. This development had complicated our lives immensely. I say development. There'd been no developing.

I'd met her over a long weekend in Pie de la Cuesta, a virgin spit of powdery beach outside Acapulco. Anne was there with a large group of people, some of whom knew a friend I'd brought along. That night we all ate pizza together at a beachside café, pulling together two long, wooden tables a stone's pitch from the Pacific. Anne sat at one end of the tables, tanned, twenty-three years old, climbing from conversation to conversation like a carefree hobo between boxcars. She kept both hands always in view above the table in the educated French manner, but when she drank she planted her elbows like a fine, deserving country girl. I sat two seats away, flirting with her when I could and as best I could. Her radiance lured me like a bug to the glow of an electric zapper. Between us sat her mother and father, visiting from France.

At some point I mentioned that I was American. The three of them sat forward slightly, in unison.

"American?"

"Is your president as ignorant as he seems?" Anne asked.

"I think so," I said, "but I've never met him. In fact I haven't lived in the States since before he was elected."

They nodded gravely.

"Is that because of your politics?" Anne asked.

"No," I said. "Just a happy coincidence."

"I can't stand politics," Anne said.

"Nor can I," I said. "It's the dirtiest business in the world."

"The dirtiest?"

"Yes."

"Dirtier than drug dealing?"

"It's different," I said. "Drug dealers don't pretend to act for the greater good. For all of their terribleness, maybe there's less hypocrisy. Also, they command smaller armies. At least they used to."

"Do you have to cover your drug war?" Anne asked.

"Indeed I do," I said. "Mostly I report on it from under my bed."

"How very brave of you."

"Not everyone can be on the front lines," I said.

"But you venture out to write about other things, I imagine."

"Once I made it as far as my street corner."

"And how does an agoraphobic make it all the way to Mexico?"

"He is still marveling at that."

I was indeed marveling. But then someone was asking about her husband's business. The bug hit the zapper. Her husband?

He was sitting two seats away from me to the other side. The quiet fellow with the goatee. Guillaume.

"I'm French too," he'd told me earlier on the beach, "but I was born in Peru." From his years in Latin America Guillaume could speak Spanish with just about any regional accent, from high Andes to Argentine *porteño*. This was a gift he possessed, as was—I was now learning—his wife. Anne.

Aghast, I confirmed the rings on their fingers with a couple of self-conscious glances. My throat twisted like a towel. How had I not noticed? I slammed on the brakes then. I could keep this cart from tipping. I was now thirty-five years old and had grown used to meeting younger women long committed to younger men. I was friends with many such couples and hastily resigned myself to the same sort of relationship, at best, with Anne and Guillaume. But most likely, I thought, we'd never see each other again.

But we did see each other. Because Anne sent me emails. Then she called.

"We're having a dinner party," she said one day, "and we need an American to pick on."

"I'm glad to help."

For my part I never said no. Though I tried my best to keep my feelings deeply buried. I mean buried in the way you bury toxic waste. And then one day not long after my phone rang again. It was Guillaume. Oh no, I thought, my storage site is leaking.

"Hey, *gringo*," he said. "We're going back to Pie de la Cuesta this weekend. You free? It'll be like last time."

"Sure," I said, relieved. "I know a restaurant there that serves freedom fries."

Around four in the morning that Saturday night Anne and I became, by a process of fatigue-driven natural selection, the last two people awake on the wooden deck of our café overlooking the beach. Everything brightly illuminated by the moon's wheel. We drank warmish beer and smoked cigarette after cigarette until our voices were hoarse and our heads were dropping. I was lighting one after another, hoping with each to delay for some small moment Anne's departure to bed.

"You know I nearly became a monk," I was saying, as a midtempo Brazilian love song began playing on Anne's portable CD player.

"Hey, stand up. Come on."

She stood and held out her hand. "*Vas-y*," she said. "Come on. Dance with me."

She was wearing a billowy orange linen shirt over her bikini top. I looked back toward the darkened bungalows, under palm trees and unlit by the moon. Guillaume had gone to bed hours ago, one of the first. As we moved in circles, my hand now on her lower back, I was thinking this is not how a married couple's new friend ought to behave. This is not how a married woman behaves.

When the song was over we slumped back down in our plastic chairs.

"*Ojo*," I croaked, pointing. "Watch out. Your chair's about to slip off the edge of the deck." Anne was sitting within an inch of an uncomfortable if not dangerous fall to the beach below but she did not look down to assess that minimal distance. Nor did she scoot forward.

Instead she just looked at me, slightly more serious now, and responded in her own hoarse voice, "*¿Quieres que te diga algo?*"

"Yes, tell me something," I said. "Why not. If you've got the voice for it."

"I'm going to tell you then. Try to remember tomorrow. *Je vois le monde à travers tes yeux.*"

The phrase caught me completely off guard. For the French, and for the meaning itself. For all of the possibilities flushed suddenly like partridges into the air before us.

I see the world through your eyes.

By the time of Lazaro's final breakdown Anne and I were blocked in an unbearable exchange of platonic longing.

"This is insanity," I told her.

"*Je t'aime.*"

"Me too. Since the very first time you insulted my Americanness."

"We have to stop seeing each other."

"Will you be my French teacher?" I asked.

That brought her to my house twice a week.

"Okay, your vocabulary words of the day."

"Okay, *je suis prête.*"

"*Prêt.*"

"Right. Ready."

"*Officiellement.*"

"Officially!"

"Correct."

"*Officieusement.*"

"No idea."

"Because it doesn't exist in English. *Officieusement* describes what's really happening. Behind the publicly accepted, *officielle* version of events."

"I need an example," I said.

"*Officiellement* we are student and teacher."

"And *officieusement* we're a couple of pathetic idiots who should be running in opposite directions?"

"You've grasped the essence of it."

"You doing anything Wednesday night? You and Guillaume of course. I host a cinema night here at the house."

So it was that Anne was at my house the night of Lazaro's snap. Guillaume had not come. If he had, Anne might not have ended up in the rain trying to talk some sense into my friend. As things stood I wasn't sure whether it was working. Lazaro was on his feet again, waving his arms about like an incensed driver after a fender bender. It occurred to me then to make some phone calls. I ran up to my office and dug up the cell phone number of Sara, Lazaro's girlfriend back in Guatemala, the newspaper reporter.

"Sara," I said, "It's Gerry in Mexico. I need your help. Lazaro has gone really crazy. He's out in the street in the rain. He won't come inside. He's in a complete rage."

"What?" she said. "What are you talking about?"

"He's in bad shape," I said.

She was silent. Then she said, "Is he under a lot of pressure?"

"I'd call these last few months 'under pressure,'" I said.

"I see," she said, pausing. "Because he tends to get violent when he's under a lot of pressure."

"Violent?"

"He doesn't handle stress well."

"He seemed to handle it okay in Guatemala," I said.

"Not always," she said.

"Listen, you've got to talk to him."

"I'm not sure that's a good idea," she said. "We've split up. I'm sure he doesn't want to talk to me."

"But you've seen this kind of behavior before?" I asked.

"Yes I have," she said.

"What should I do?"

"I don't know. Ask him."

I hung up and called Susana.

"Oh no," she said.

"I don't know what to do," I told her. Then, "Why did you say 'Oh no' like that?"

"I never wanted to tell you this," she said, "because I wanted Lazaro to have a second chance, a clean second chance."

Oh shit, I thought. "Tell me what, Susana?"

"You knew Lazaro and I had split up. Do you know why? I have a restraining order against him. From a judge. He became obsessed with me. When I ended our relationship he came over one night and tried to break into my house. He was totally crazy. I had to call the police. After that I went to court and got the restraining order."

"When did that happen?" I asked.

"Last year. I hadn't seen him in months until he burst into the office that day saying that someone had tried to kill him. In fact, he wasn't allowed in our offices either."

I looked outside. Anne and Lazaro were still at the curb.

"Why not?" I asked.

"We suspected he was stealing from us," she said. "To pay for his drug habit."

"What!?"

"He spent all of his money on cocaine," she said. "It was really bad. You should have seen where he was living. His apartment was so dirty and barren. But I knew he was trying to clean himself up. And then, this."

"A drug problem?"

For a moment I stopped listening down the line. As Susana talked, the entire last three months, from the morning of the kidnapping to this chaotic moment, were undergoing a rapid and radical revision.

"The story missed for the details," I said aloud. "What if all the signals I've been attributing to posttraumatic stress are in reality the symptoms of someone strung out on coke? The nervousness, the fatigue, the depression, the unexplained disappearance of money in the house? My pants? Now at least I understand why everyone seemed so reluctant to help Lazaro on the day of his abduction. You'd already banished him from your lives. Maybe Lazaro's abduction had nothing to do with my 'materials' and everything to do with some drug debt he's accrued. The 'military-looking' men might have been some drug boss's muscle sent out to collect on that debt. But what could they collect from a poor kid like Lazaro? His head, of course. But even better? His truck. Maybe they'd taken the pickup as payment and thrown Lazaro into the street, just some hopeless Guatemalan nobody, not even worth the lead in their bullets. The truck, after all, has never been found. Not a trace of it. It was easily worth more than ten grand. Maybe Lazaro then deliberately connected the kidnapping to my work to cover up his personal predicament—and at the same time in order to get himself a free ticket out of Guatemala and away from the reach of men notoriously inflexible when it comes to things like failure to pay for product."

"Gerry?" Susana said. "You're losing me."

"I think I'm lost myself," I said. "I'll call you back tomorrow." I hung up and ran downstairs just as Anne came inside. Lazaro was still sitting on the curb down the block. It was still raining.

"He's very upset with you," she said. "That you changed the locks."

"Please tell Lazaro he can come back in to get his things, then he's gone. And only with an escort." I'd already called Marion Lloyd and asked her to come over. She and her Mexican fiancé, Chicharro, accompanied Lazaro to his room. I waited by the front entrance. A few minutes later they came down. Lazaro looked like some surrendered hostage taker being escorted by police from a bank. He kept his head up and his gaze straight in front of him, flanked by Marion and Chich. I started to speak, but Marion held up a hand to silence me. Then she made an 'I'll call you later' gesture by holding her hand to her ear and they were gone. I closed the door and went upstairs. Everyone was standing around the living room. Some already had their coats on.

"Does this mean we watch a different movie next week?" someone asked, "or do we continue where this one left off?"

The next morning I woke up angry and anxious to get to the bottom of things. I called a Guatemalan newsman, an editor named Eric who had given Lazaro his first assignment as a photographer several years earlier. Eric was one of Lazaro's mentors.

"I had lunch with Lazaro just before your last trip to Guatemala," he said down the line. "And he told me that he had just returned from a month at a drug rehab center."

"Had you known about any drug problem?" I asked.

"Not until that moment," he said. "But I hadn't seen Lazaro for a long time."

I hung up, no closer to understanding what had really happened. I'd never sensed Lazaro was wired on blow. And I'd been around the stuff enough to know what the signs were. Maybe Lazaro had been clean during my trips to Guatemala, even my last one. But then the stress of fleeing his homeland pushed him to the edge and he'd relapsed, seeking out cocaine again in the crowded streets of Mexico City. And if that were true, he would have been under intense pressure to come up with money to support his habit.

After lunch I went to his room and went through the garbage and the few belongings he'd left behind. I found no signs of drug use, but I did find a telephone number for his sisters. I dialed it. A woman answered and when I enquired if Lazaro was staying with them she said flatly, "We have no more contact with him," and hung up.

I thought of Ambassador Bushnell and her extraordinary gesture that probably had saved Lazaro's life. And NPR was still sending me extra money for Lazaro each month. What could I tell them now? Thanks, everyone, but there's been a misunderstanding. You didn't just save a courageous young Guatemalan journalist nearly stamped out by a corrupt, narco-oligarchy posing as a democracy. You saved a lost young Guatemalan deeply mixed up in the same world of drugs he was helping us to report on.

Or maybe this was not the correct line of reasoning. Maybe drugs were completely unrelated to the kidnapping. Lazaro's Guatemalan contacts were convinced it was a drug issue, but colleagues in Mexico who knew Lazaro had trouble buying such a story. As I mulled over his three months here in exile I couldn't discard any version of events. I was simply unsure. The answers lay with Lazaro, but I no longer knew where to find him, and for the moment I had no more energy to put toward the mystery. The ordeal had left me exhausted and shaken. I went up to my roof and fell asleep in my hammock. When I woke up it was dark.

A week later I met up with Marion Lloyd. She gave me a scant update. "Chich and I put Lazaro up in a youth hostel that first night, partly because we were afraid to have him in our place, given his highly agitated state. I called the next day," she said, "but he'd checked out. I haven't heard from him since."

Walking home through the Condesa, I asked myself if I'd done the right thing after all. I'd been wondering this since the moment the locksmith had arrived. But now something larger was weighing on me. Lazaro was the fourth person I'd booted from the house on Jojutla Street, if you counted Concha and her two sons. A man needs his own space, I'd told her younger boy, Alejandro, one day under my stairs. It wasn't just that I was feeling more and more lonely in my spacious residence. This was one contradiction, but not the most disturbing one. As I inserted the key into my new front door lock I was remembering how Buddhists were supposed to try to help people—all people. In my own ragged, deteriorating practice I'd come to be working for only a select, and in some ways easier, group. I'd become adept at beseeching troublesome ghosts to stay and seek peace in the house. But when real people became troublesome I showed them the door.

As I stood in front of the hole busted through the jailhouse wall I was wondering how you might patch something like this, or if that was even possible now. Literally every prisoner had escaped. And among those now on the lam were a handful of very angry former Aristide supporters—the same supporters who'd been helping him hang onto power.

Aristide had been relying on a patchwork of local gangs, from the slums of Port-au-Prince's Cité Soleil to the inland hamlets and coastal cities such as this one—Gonaives—to do his bidding. In exchange for money and freedom to carry out their criminal enterprises, these gangs provided Aristide with what amounted to protection. With a phone call from the National Palace, Aristide's underworld clients would mobilize their forces to march in his support. Or they would stream out into the streets to dissuade anyone bent on protesting against the government.

Aristide always denied having relations with these shadowy figures, known locally as *chimères*. But strong evidence to the contrary had just run for cover from this jail: his name was Amiot "Cubain" Metayer. Cubain ran a depraved local criminal gang called the Cannibal Army. Its ranks swarmed with thieves, thugs, and former soldiers, and it controlled the lucrative trade in contraband here through the port in Gonaives, a three-hour drive north from Port-au-Prince. I'd come up here to talk to him.

But it was unclear whether Cubain was in a talking mood. He'd had a major falling-out with his president. The problem started because while Aristide might have been willing to give this local gangster a wide berth to run his illegal rackets, the United States was not. It wanted Cubain in jail. About 15 percent of the cocaine entering the United States passed through Haiti, and American investigators accused Cubain of having a hand in it.

Aristide eventually bowed to American pressure and threw Cubain in the boot. But unluckily for the president, Cubain wasn't just another gangster and Gonaives wasn't just another city. It was the revolutionary heart of Haiti, the place from which rebel-slave Jean-Jacques Dessalines first declared the country's independence from France on January 1, 1804.

When word hit the streets that Aristide had thrown Cubain in the clink, the Cannibal Army rioted for several days, burning down the customs house and generally ransacking things. Then Cubain's brother, Butter, laid siege to the jail head on.

"All they did was, like, keep shooting nonstop," a local hotel owner told me. His hotel was just across the street from the jail. Its entrance had swinging saloon doors. We walked through them and then up a long wooden staircase so that he could show me where he'd lain on the floor for protection until the shooting was over.

"And the whole time they're doing this, there's another group, with a front-loader. The guys in the front, they're keeping the cops occupied while from the back all you could hear was this Boom! Boom! Boom! It was like watching a movie. I think the only thing they wanted was to get Cubain free."

I had driven up to Gonaives with veteran Haiti reporter Michael Deibert, then the Reuters correspondent in Haiti. Deibert had been on the ground long enough to have cultivated some good contacts in Gonaives and, after interviewing the local hotelier and checking out the hole in the jail wall for ourselves, we sat down with one such acquaintance at a curbside bar for a round of beers. Around us, life seemed to have returned to normal, though the dusty streets did still harbor the remnants of myriad unidentifiable objects charred during the recent unrest.

Our contact—a tall, skinny man wearing a black leather vest with no shirt underneath it—sat silently drinking beer after beer. So we did the same. As the evening wore on and the collection of empty bottles on our table grew, Deibert's contact abruptly made the move we'd been counting on.

"Let's go to another bar," he said, "in Rabateau." Rabateau

was a seaside neighborhood of Gonaives. It was currently barricaded owing to the unrest—and it was where Cubain was likely holed up. It was not a quarter through which one took leisurely nighttime strolls.

We piled into the man's car and puttered past more blackened evidence of the fires of protest until we reached a road lined with expensive, imported four-by-fours, BMWs, and pickups. They were parked fender to fender for more than a block and on the roof of each sat a man with a rifle or a shotgun. There were no functioning streetlights but there was a glow from the bar at the end of the road. We parked in front of it and climbed to the sidewalk. I was feeling pleasantly unsteady from the heat and the drink. A big sign above the door of the club read CHANDEL. We went in.

Inside the Chandel it was gangster heaven and something like 30 degrees colder than on the street. The joint was lit with fluorescent bulbs that caused the white parts of your clothes to glow annoyingly. The *compa* music thumped throughout that gaudy, mirrored lair of tough guys, and we joined a table of extra-serious-looking specimens, all wearing black hats, their arms around doe-eyed women with whom you dared not make eye contact yourself. Just as at the street corner bar earlier, the conversation with our hosts ran from zero to nonexistent. No room for idle banter among idle killers.

Occasionally I caught a glimpse of guns tucked into men's belts as they got up to dance or sat down. When they danced it was with gorgeous, long-legged women, lighter skinned and probably Dominican, whose gyrating hips spoke to how their nights would end. This was the glorious reward for the Cannibal Army's upper management.

There was a long period of hopeful waiting in which Cubain might have appeared, but he did not.

The next day we tried again to meet with the gangster. Heads pounding, the sun pounding, we returned to Rabateau with Deibert's contact, climbing by foot this time past a barricade of debris and across a sewage-stained concrete road that marked the outer frontier of the Cannibal Army's jurisdiction.

A short stroll later we came to a small cinderblock house under construction where several gunmen lounged in what shade they could find. We waited for a long time and were finally told that Cubain would not emerge to speak with us. The situation was too unstable. Instead, we chatted with one of his fearsome lieutenants, a stout gunner named Winter Etienne. Etienne was overweight and had wild hair and seemed as intelligent as he was vicious.

"The Haitian people have now seen how this government is doing things," he told us. "And we see that it's the same as under the Duvalier dictatorships. People can't eat. Life is still really hard in Haiti. We want this situation to change."

Indeed, as many had foreseen, Haiti's economy had remained in the dumps over the year and a half since Aristide was sworn into office. International aid was still on hold, allegedly until the contested May 2000 legislative vote could either be rerun or recounted. And with each passing week the chance of that happening grew dimmer. The National Assembly also remained paralyzed, making it impossible for the government to enact new economic stimulus packages.

"Cubain is one of the people trying to save this country," Etienne went on, defending the jailbreak. "Like in the Bible, Joseph was accused by his brothers and thrown into jail despite being innocent. This is what happened to Cubain. And it's why we are calling on Aristide to leave Haiti. He deliberately arrested an innocent man."

Taking Etienne's bold demand at face value, it appeared that Aristide had, in a matter of days, turned a powerful ally against himself and lost control of one of the country's most strategic and symbolic cities. He might have taken solace, momentarily, from the fact that the Cannibal Army's call to rebellion was only resonating locally. But two more miscalculations that same summer greatly weakened the president's support among Haiti's young and vast populace of poor across the country.

First, the president tried to replace the well-regarded rector of the State University of Haiti, located in Port-au-Prince, with a crony. The move kindled a brush fire of student protests that he was now finding difficult to stomp out. Second, a homegrown investment project involving local cooperatives, championed by Aristide himself as a way to generate wealth without outside assistance, collapsed. It turned out to be a scam—a $200 million pyramid scheme in which an untold number of Haitians lost their life savings. Among the victims was a middle-aged charcoal seller whom I met

on the outskirts of Port-au-Prince after returning from Gonaives.

"It was all of my resources," the man told me sadly. His
skinny arms were black with soot to his elbows. He waved
them like two charred sticks. Before him, rows of white sacks
stuffed with Haiti's sole source of cooking fuel. "Everything
I had is gone!" he said, holding out a handful of coal as if I
might divine in it his misfortune and his prospects. "I have
four kids and a wife! I have no hope. Everything is totally
lost. This year my children cannot go to school because what
we had, they took it."

Aristide had since come forth and assured the public that
the government would cover investor losses. But the charcoal vender said he suspected Aristide's promise was no
more than a ruse to calm the people down.

Ruse or not, it only worked for a little while. Soon enough
the Cannibal Army was on the march toward the capital.

On my flight home from Haiti I lay back and tried to reflect on all that I'd seen and heard while there. But I couldn't get into it. I kept returning to what I hadn't seen or heard. To Anne. I was glum because I knew I wasn't going to see her for several more days; she and Guillaume had gone to Miami to see his father. I was imagining my big Mexican house, how much bigger it was going to feel with Anne so far away. With Anne still together with Guillaume.

"I can't leave him," she'd told me before I'd left for Haiti. "But I love you."

We had never even kissed, and yet I felt deeply committed to her. To the ridiculous idea of being with her. It was shockingly apparent that I needed to yank this weed out at the root. Three months had now passed, and the only things I had to show for it were bags under my eyes and a folder full of agonizing emails. Plus some French.

As my plane came in for its landing I resolved to tell Anne
that our unborn relationship was finished. She needed to
resolve her marriage, for better or worse. If you find yourself
single one day, I was imagining myself telling her, that might
be another story. But for now it's time to move on.

By the time I'd climbed into a taxi at the airport I was
feeling remarkably better. What a fool I'd been. Then the
taxi pulled up in front of my house and my heart jumped. It
was seven in the evening and the lights were blazing on the
second floor, the main floor, of Jojutla 13. Walter and Guadalupe, the Argentine friends staying with me, normally
worked until late. I wasn't expecting anyone else at that
hour. And no one else had the key. Except Anne. That had
been one of my gestures. Might her trip have been canceled?
Might *she* have canceled it?

I fumbled with the change from the cabbie then unlocked
the garage door and ran up the front steps into the living
room.

"*Hola,*" I said, panting, my adrenaline booming. I set
down my bags.

She was sitting on the couch, huddled against one corner
of it, clutching a large cushion to her chest. The TV was on,
as was the radio and every single light on that floor. When
she saw me she leapt up and ran across the room and literally
jumped into my arms.

"*Tranquila,*" I said, stumbling backward under her weight.
"*¿Qué pasa, Karina?*"

Karina was a young Mexican woman from Acapulco, a

friend of Guadalupe and Walter's whom I was letting crash at the house for a while. In Acapulco they called Karina "The Panther" for her long dark legs and her predatory smile. I'd completely forgotten she was here; Jojutla 13 was becoming something of a mix between a refugee center and a roadside inn. Karina and I barely knew each other—certainly not well enough for me to merit such an impassioned reception—so I repeated the question.

"*Qué pasa?*"

"Excuse me for jumping into your arms like this," she said. Her femme fatale smile was gone. She looked more like a little girl now. "I'm just a bit scared. I was sitting here watching TV when a young boy came down the stairs. He stopped there on the landing as if he were looking at me. But he was only pants so I cannot be sure."

"Only pants?"

"Yes, he was wearing knickers to his knees and long socks and brown shoes. From his waist up there was nothing. No clothes, no body. Like he'd been cut in half. He turned and came down the stairs and went past the dining room then into the kitchen."

"Into the kitchen."

"Yes."

"Is he still there?"

"I don't know. That was, like, four hours ago."

"You've been huddled on the couch for four hours?"

"Maybe five."

"May I put you down?"

"Yes, of course. If you want to. You don't have to."

"That's okay. Why don't we go check out the kitchen together?" I said. Karina's response was to run back to the couch and take up her defensive position behind the cushion.

"All right, so you wait here," I said. I approached the kitchen, tracing the route Karina had described, then swung the door open. I walked in slowly, holding the door with one hand. The lights were not on here, so I flipped the switch.

But I found nothing out of the ordinary.

Pants Boy must have already pattered away.

I glanced toward the eerie back stairwell that wound directly up to my office and thought, like hell I'm taking that route tonight. Then there was a tremendous slamming sound from down below, followed by footsteps coming up the stairs. Karina screamed. I ran back to the living room.

"*¿Qué onda?*" Walter said. *What's up?* He was still wearing his motorcycle helmet and he looked like a spaceman.

"*Hola*," I said. Then, "Karina saw Pants Boy."

"What?"

"There was a pair of pants that came down the stairs," she said.

"We need to talk," Walter said, shaking his helmeted head. "Tonight. It is worse than I thought."

"They weren't your pants," I said.

"We need to talk about this house," he said again. "About how to fix this house."

"Would you mind taking your helmet off?"

"*¡Sí!*" shouted Karina. "Let's fix the house!"

"Would you two stop?" I said.

"We brought in a shaman to cleanse the restaurant to-night," Walter said, sitting down on the couch, setting his motorcycle helmet on the floor. He rubbed his salt-and-pepper chin beard. Walter was helping an American inves-tor get a trendy new fish joint off the ground. Bringing in a shaman to clear bad vibes is not unusual when starting a business in Mexico.

"There were, like, thirty of us in the restaurant," Walter went on. "And this old Indian guy, he walks in and without even glancing at the sushi bar, which is really gorgeous, he walks right up to me and asks, 'Where do *you* live?' I tell him, with a friend. 'Not with whom—where?' he says. In the Condesa, I tell him. In a house. In Mexico City. He asks me if it's haunted. I ask him why and he says, 'Because there's a human soul on your back.'" Walter paused then for effect. "'And it's sucking out your life energy.'"

Karina and I looked at each other. Her eyebrows were raised like mine. We circled behind Walter. I couldn't see anything unusual. Karina touched the back of his neck with one finger, then retracted it.

"You feel anything?" I said.

"I don't know."

"What do you mean, you don't know?"

"I *have* been feeling tired lately," Walter said. And that's more or less when the same idea dawned on us all. From my first days here it had been obvious that Jojutla 13 was caught up in some sort of spectral malaise. At least since I'd arrived, but probably longer. And by extension anyone who lived in the house was caught up in it too. The time had come now to break the spell. It would require drastic measures—something far less workaday, far more profound, than the pre-Colombian incantations of a shaman.

"Let's give Jojutla 13 nothing less than a complete interior design makeover," Walter said. "It's bound to work."

"That's it," I said. "The less scary the house looks, the less reason they'll have to haunt it!" We knew where to start: by painting over the mottled copper walls and ripping out the billiard-table-green, wall-to-wall carpeting, which everyone agreed had only served to make the unseen feel more at home.

I'd been in the house for more than two years now and had grown used to my light-absorbing chambers, my Mexican bourgeois black hole. So I was surprised by how much I appreciated the change unfolding before my eyes.

"Looks great, everyone! You're working miracles! I shall repay you all tenfold in mescal!"

A crew of friends and I were steadfastly applying a coat of decidedly less spooky off-white to the walls while another group tore at the mangy carpeting. It lifted easily to reveal a treasure: a pristine floor of square tiles, terra-cotta, polished smooth and interspersed with small white porcelain pieces painted with human figures in blue. Darkness be gone, I thought. Ghosts be gone. I already had more than enough uncertainty in my life without having to worry about creatures I couldn't even see and who seemed bent only on freaking everybody out. I glanced at Anne, back now from Florida, roller in hand, her face speckled with paint like a robin's egg.

The phone began ringing in the office. I put down my brush and sprinted up the stairs.

"Gerry, Didi."

"What's up?"

"U.S.-Mexico relations just disappeared."

"How can relations disappear?"

"Texas just executed Javier Suarez Medina."

"I don't believe it."

"Log on," Didi said. "Become a believer."

Fourteen years and one day earlier, Suarez had machine-gunned to death an undercover Dallas police officer in a drug deal run mightily off course. And now, I was reading, he was dead, executed in Huntsville by lethal injection.

Suarez's execution didn't create much of a stir stateside, but from what I could see online it was already provoking a holy, indignant outcry here. This was to be expected, because in killing him, the Americans violated the terms of an international treaty giving foreign nationals the right to notify their embassy or consulate of their arrest. Suarez was never told of this right. When a jury sentenced him to death, millions of Mexicans clamored for a new trial.

It never came.

The Mexican press demanded that his sentence be commuted to life in prison. Even President Vicente Fox Quesada tried to save him, at the last minute. Fox was anxious to capitalize on such strong public sentiment. In theory he was to be at Bush's ranch, in Crawford, Texas, the day after Suarez's execution date—and he had no intention of showing up to collect a casket.

Fox called Texas governor Rick Perry but was told that justice must take its course. So Fox went over his head. He rang up President Bush and pleaded for a presidential pardon. It was the least Bush could do. He'd let Fox down on every other occasion since the two had taken office. Don't kill this Mexican, Fox asked. Not this one.

Now, with Suarez flat-lined, Fox was at home unpacking his bags. Outraged, and without an ounce of political capital left to spend on Bush, he had ordered his presidential plane back to its hangar. His horse ride with "Jorge" was on hold, indefinitely.

By the time I'd finished getting the story together and filing for the next day's *Morning Edition* it was after ten o'clock. I went back downstairs but everyone had gone. Anne was gone too, her paint roller cleaned and drying with the rest of them in the kitchen. My house was brighter now, but it was still too big, I thought. There was no getting around it. I ate some leftovers there in the kitchen then headed for the staircase to go to bed. Then I saw the note, painted in French in big letters on a part of the living room wall still copper-colored. And the house shrank a little.

À demain.

See you tomorrow.

With what stealth the first anniversary of the September 11 attacks came upon me. Had a year already passed since I'd been sitting tangled in my Bogotá bedsheets? It was true that the world had changed, I thought. I was now kissing Anne for the first time, on the floor in my dining room. The tiles beneath us were cool and smooth, and I found myself longing unexpectedly for the softer padding of the old green carpet whose removal had either coincided with or helped end the increasingly unpleasant ghostly surfacings.

"Leave him," I was saying. "You know, you've never taught me to say that in French."

"He wants us to move to Lerida, on the Yucatan. He wants to start a family."

My gut knotted.

"But you don't want that."

"No. But it's harder than you think. I married him. People don't just throw that away."

"But what is the 'that'?" I asked.

"What?"

"What would you be throwing away? A marriage that isn't working? Is that something you think people should keep?"

"You've never been married," Anne said. "How could you understand such a commitment?"

We'd had this conversation now a hundred times. Now that we'd kissed I wanted her more than ever and we were as stuck as ever.

"I understand a few things," I said. "Among them is that we've just crossed a line. If Guillaume catches us. Imagine. This can't go any further. You've got to act quickly. I don't want to get caught up in some giant mess."

"Give me time," Anne pleaded. Her hands started to tremble in mine. I backed off.

"I can give you time," I assured her. "Hey, I'll give you time."

What other choice did I have? Her hands were shaking, but mine were tied. She was in a terrible fix, really. Twenty-three years old, recently married, halfway around the world. Put yourself in her shoes, I told myself. She came with one man but has sought out another she barely knows, and he comes with a wrecking ball.

Anne left that afternoon visibly upset.

"I don't know how I'm going to hide this at home," she said, walking out the door.

"Then don't hide it!" I called after her, unable to resist.

I went up to my office to try to lose myself in my work. It

was going to be easier than I'd thought. There was an email waiting for me from foreign editor Loren Jenkins. He was sending our South America correspondent, Martin Kaste, and me on a program known as hostile environment training. HET entailed running around the Virginia woods for a week with a bunch of former Special Forces commandos, learning how not to get your backside shot off while reporting. The term "hostile environment" struck me as vague. It might be applied to anything from urban street clashes to rural unrest to outright warfare. Reporters in Latin America generally see their fair share of the first two scenarios. But given that no such reporter from NPR had ever been sent to this sort of training before, I could only conclude that Jenkins was readying us for something new. He was prepping us for the war in Afghanistan and the specter of it in Iraq.

We were riding on an old school bus through the countryside when two roadside bombs flashed and exploded just ahead of us. The bus driver slammed on the brakes. People were screaming. Then three hooded men stormed the bus with automatic weapons.

"Get your goddamn fucking heads down! Now! Get your heads down or I will blow them off, media scum!"

The men ran up and down the aisle tying canvas bags over our heads.

"Don't anybody fucking move!" We didn't. We sat still in our seats for long minutes. I could hear someone pacing up and down the aisle. Then whoever it was cuffed me on the back of the head.

"You," he said. "Stand up."

"Me?"

"Shut the fuck up!"

He dragged me off the bus and out into the cold air and the snow.

"Get down on your fucking knees, scumbag. Hands behind your back."

"Please," I said. "I won't make any more trouble."

"What were you doing then, fiddling with your hood, lying scum journalist?"

"I couldn't breathe."

"You don't think the rest of your asshole companions were suffocating? You had to go and stand out, huh? Draw attention to yourself? Cause me to get bloody fucking angry?"

With that the man lifted the bag off my head. The rest of the journalists from the bus were standing in a semicircle in front of me.

"Okay, listen up," my aggressor said in a loud voice. "Rule number one in a hostage situation: Don't stand out. Become the 'gray man.' Kidnappers don't take kindly to troublemakers. Kidnappers are most likely to kill troublemakers than quiet captives. You can stand up, Gerry. But if this had a been real situation, you might not have gotten up again."

I considered that a pretty good thing to learn. This hostile environment training is really useful, I thought. The next day we went on a hike and were shot at. After running in circles like scared rabbits, we were gathered together in a circle by our instructor.

"If you hear shots," he said, in a heavy northern English accent, "you hit the floor. Look for a depression in the earth. Don't move. If you get up to run before you know where the gunman is, you're dead. Also, by the time you hear the sound of a gun firing, the bullet is already past you. So that's an indication that you're not dead."

It seemed there was no end to what we could learn. For basic first aid we were taught how to make ad hoc stretchers and splints. How to staunch bleeding. In one role-play we came upon a car that had hit a roadside mine. I had to apply a tourniquet to a man wearing a latex stump over his hand simulating an amputation. Beneath the latex he held a little squeeze bottle filled with fake blood that he squirted in my face as he writhed and bellowed in the snow. "Ay, me fuckin' hand is gone! Where's me bloody fuckin' hand?!"

For urban conflicts, we were advised to scout out protest scenes beforehand to predetermine escape routes should things get nasty. Also, we were instructed never to place ourselves between riot police and rioters. If you do, there's a good chance both sides will end up thumping on you. That one seemed like a no-brainer, but we watched real news footage of people who'd made that very mistake. Their violent deaths drove the point home.

"Do you think we're going to Afghanistan?" I asked Kaste on our last night.

"Seems inevitable," he said. "Everyone's rotating through."

But I wasn't going to need an all-out war to test my new survival skills. Because the very next day Jenkins dispatched me to Venezuela. I was covering for Kaste, who had a family emergency. In Caracas, millions of people were now marching against President Hugo Chávez, and we needed someone there to cover it.

I arrived feeling as safe as a hockey goalie in full pads. On my first foray I went to cover a massive confrontation between pro- and anti-Chávez crowds. When the insults began flying across a police line separating the two sides, I still felt pretty okay. But when the insults became bottles and the police got riled, I surprised myself by promptly forgetting all of my fresh training.

I ran from one side of the elevated plaza to the other as the cops fired round after round of teargas at what I could have sworn was me. Then I managed to get myself caught in an ambush as hundreds of Chávez goons snuck around behind the square where all this deep stupidity was unfolding. One tossed a powerful firecracker at me as I was running toward a railing—the escape route that I'd selected only then, on the fly. The device blew up behind my head as I was hopping over the waist-high metal bars. I fell forward, lost my footing, and crashed to the dirt another six feet below. For a moment I couldn't move, which I judge now as a positive turn of events because a steady stream of ambushers leapt over me, swift as wolves, without as much as glancing down at me. I have sought out this depression and I am lying in it, I lied to myself, just as the commandos had taught me. Then the teargas drifted my way and I was on my feet again, hobbling along with a crowd of people down a narrow street. It was hard to see with my eyes so watery. Everybody was yelling. I felt like a bull at Pamplona.

Three weeks later things hadn't changed at all.

"Anne? Anne, it's me! Can you hear me okay?"

"Are you back?"

"No. No, I'm still here in Caracas. I'm under a car."

On the street around me wackos loyal to the government and opposition psychopaths were exchanging volleys of rocks with zeal. Then someone had started shooting from an apartment window.

"It's been three weeks. When are you coming?"

"I hope I can be home for Christmas," I said. "Did you tell him yet?"

There was a moment of silence.

"No, not yet. Why are you under a car?"

"Anne," I said, just as something, a bullet or a big rock maybe, smashed the windshield above me, "the sooner you do it, the better. I don't think the perfect moment is ever going to arrive."

Just before Christmas it had become clear that there was nothing new to report. Chávez was still in power; millions still wanted him out. So my editors sent me home. On my return to Mexico City I bought a bouquet of bottle rockets from a roadside stand, inspired by the enormous quantities of gunpowder that had exploded around me over the last month. I led Anne up to the roof. She should have been at home, cooking Christmas Eve dinner. We didn't have much time.

"Merry Christmas," I said. "Close your eyes." When the

first rocket fuse began to sputter Anne opened her eyes. Up it went. I readied for a big explosion—loud enough—convincing enough, I hoped—to herald a sort of turning point in this madness. But what we heard was more of a pop, guppy-sized, without resonance and swallowed up quickly by the city's endless predatory white noise.

"I'm sorry you can't come to dinner," Anne said, watching the tiny cloud of black smoke dissipate over the Chapultepec Castle.

"Me too," I said, "Everyone's going to be there."

"But it would be too uncomfortable for both of us," she said.

"I know."

"No, you don't. Not entirely," she said.

"What?"

"I've told him."

"You did?" I said. "Really? When?"

"Yesterday."

"I had no idea," I said, "That's fantastic." Thank you, bottle rocket, I thought excitedly. "So? How did he take it?"

"I told him that I'd fallen in love with someone else. And he knew who it was."

"He did. Jesus. I knew it. What did he say?"

"He's a problem solver," Anne said, rolling her eyes in exasperation. "He had a solution prepared." She put her hands over her face. "He's asked me never to see you again. He begged me. He thinks that will fix us."

"But how can you fix the fact that you don't love him?"

"I just don't love him the way I used to."

"You know what I mean."

Anne was fiddling with an unlit rocket. The silence felt dank and sad then. Stripped of mystery.

"You're going to tell me you can't see me again."

"I'm so sorry," she said, "I'm so sorry."

"But the whole point was to tell him you'd fallen in love with someone else," I said, "and then to *be* with that someone else."

"I promised him. I've promised him."

"You're saying everything twice," I said, looking to buy time, for any way to change this. "Say 'I'm leaving him' twice."

For a moment it seemed to work. "*Cabrón*," she said, smiling. "*Cabrón*."

Then she leaned into me, unsmiling, and kissed me that last time. "Until another life," she said quietly. She stood and went back inside and down through the house and out onto the street. I could hear her footsteps, or somebody else's, heading down Jojutla Street.

A long time later, when the cold night air had taken temporary possession of my grief, I stood as well. I stumbled downstairs, sat on the couch, and ate something in the shadow of the Christmas tree I'd bought that very day and had yet to decorate. I had no ornaments nor thoughts of any. All through my newly repainted house not a creature stirred, not even the ghost of a mouse. I could hear myself chewing. Jesus Christ on a Harley, I thought, it is time to get out of Dodge.

I bought a ticket back to Guatemala City.

The next day as I was packing my gear someone rang the front bell. I peeked over the railing. At first I didn't recognize him, but then I did. So he'd finally come back. Somehow I knew he would. I went downstairs and stepped out onto the sidewalk.

"I wanted to talk to you," Lazaro said. He looked bad. His eyes were bloodshot and his face was drawn and half hidden beneath his moppy hair. He was leaning in close to me like a drunk who'd lost his sense of personal space and this made me nervous, as did his fists. "I didn't steal anybody's money."

I looked at him, waiting for more.

"You think I stole three hundred dollars from Guadalupe."

"That was a while ago," I said slowly, "but yes, we suspected it was you at the time."

"You know I would never steal like that."

"I never would have thought so, and I'm glad to hear you say it," I said. "But yes, you were on a short list of suspects."

"Suspects."

"What do you expect? You didn't do anything to convince us otherwise. And you wouldn't talk to me. Between your moodiness and lack of work and, on top of it all, your cocaine problem, I began to doubt you."

Lazaro didn't take the bait. "You shouldn't kick a guy when he's down," he said. "I'm trying to pull myself out of this mess. My life has been ruined."

"I had a right to talk about this with my friends," I said. "There are a lot of unanswered questions and you've been less than forthcoming. If people start gossiping I can't control that. I'm not trying to kick you while you're down. On the contrary."

Lazaro seemed to think this over. "But I was always straight with you," he said.

"I know," I said. "That's why I always called you. I trusted you completely." We were both leaning against the front of the house now, staring across the street at nothing. I was picturing the silent, smoldering anger on Lazaro's face on that launch in the Guatemalan estuary as the sky darkened, his determination to get my money back. I imagined his eager grin when he'd handed the Associated Press producer her missing microphone cable in San Salvador. The fear in his eyes as he writhed on the floor in Susana's office, one pant leg ripped open at the knee. I remembered toasting to a job well done in the Petén.

"Where the hell have you been anyway?" I asked.

"Around," he said. "I've been organizing a photo exhibit

of Guatemalan peasants. My photos. I'm going to take it around Mexico with people from an NGO that works with Central Americans."

Lazaro and his family had always been traveling around, and I imagined him adapting easily, even benefiting, from a few weeks touring the countryside. Who knew? He might even get a job with the NGO in the end.

"You're going to be okay," I said. "You'll see. Maybe one day soon we can even hit the road again, together."

"We'll see how it goes," he said, ignoring my comment. "I'll be in touch. And please tell Guadalupe I didn't take her money." He shook my hand solemnly. "And tell the others too. Tell Anne."

I nodded wearily. Between him and Anne I felt as depleted as a San Luis Potosi silver mine. I waited outside and watched Lazaro walk away, slouching somewhat and heading for who knows where. If I'm exhausted, I thought, what's he running on? As his figure grew smaller I imagined him climbing onto some bus and riding it to the end of the line. He was good at orienting himself that way. But the map he was charting now was exile and was unpromising and he was lost on the inside. I went back into the house to gather my bags and begin an exile of my own.

I spent the next several months in self-imposed perpetual motion, vanished into my work. Slingshotting from Guatemala, where the United States was scolding its leadership for sleeping with the cocaine cartels and where human rights workers were dropping like flies, to Jamaica, where AIDS was spreading like the virus it is and where being openly gay meant getting your legs busted by a society in denial. Then back to the U.S.-Mexico border to visit some of the thousands of deported minors languishing in detention centers, far from home and far from their parents in the United States. Some slipped me letters, unaddressed, because they did not know by heart where they lived or where their kin lived. I made my way to Ciudad Juárez, tomb of women. To that border city's apathetic or complicit or terror-stricken authorities. A "crimes against women" prosecutor lying to my face, his eyes like a fish's on ice, claiming to have arrested the killers. Then on to the brutal desert beyond, to the maquiladora villages ringing the tormented metropolis. The hidden multitudes living in the wind, without lights, their houses built from old truck tires and plastic sheeting. This embarrassment of NAFTA.

It was all to avoid being drawn back to Mexico City. It was hard enough being on the road and finding no sign of her. Then one morning, while wandering south of the border near El Paso, it occurred to me that the moment had long arrived to do the story I'd been planning since before 9/11: to show what crossing the border was really all about. For undocumented Latinos, that is. Especially those from Central America and points south. It wasn't just sneaking past searchlights south of Tucson and a happy reunion with family. It began with a thorough uprooting, the rending of all you knew. That was the start. Then came a panic-filled journey at the mercy of people who looked right through you, who might roll you at any moment. Strangers, gangs, cops, coyotes—nobody was to be trusted. Nobody along the way was going to care whether you made it. They were there to shake you down. Welcome to the economy of immigration. *Bienvenidos*. You are now a raw material. A commodity to be traded, transported, sold, and sometimes discarded. And yet subjecting yourself to this dehumanizing treatment is preferable to staying at home because at home there is nothing. In that sense the story began to obsess me. I think that's what drove me then to tell it.

I was struggling to keep my car moving steadily but slowly along the irregular dirt tracks that crisscrossed these lush mountains. Eastern El Salvador was coffee country. But the boom years were over. It was now the era of the global coffee glut, in which coffee prices had plummeted because of worldwide overproduction. In the fall of 2003, El Salvador's thousands of small terraced plantations no longer held the promise of fortune. This was as good a place as any to start the journey. Or so I thought.

I drove for a long time looking for a plantation with some sign of life, but I couldn't find a soul out picking beans. Over a third of El Salvadorans had already emigrated to the United States but today it seemed that the whole country had packed up and left. At midafternoon I stopped on the outskirts of a village called Berlin. Inside the local coffee cooperative, the depository where local growers sold their beans, I found its director, Roberto Durán.

"I apologize for dropping in without having called," I said.

"Not to worry," he said. "You haven't interrupted anything at all." I believed him. Durán was a big man in a white short-sleeve shirt and tie, and he led me into his office where he seated himself behind an enormous wooden desk with nothing on it. No pen, no piles of paper. Not even a telephone.

"I've been driving around a lot," I said, "and I haven't seen anyone working."

"Because coffee prices are too low," he said matter-of-factly. He was rolling a single green coffee bean between two fingers. "It now costs more to pick and process the beans than to just leave them. More than half of the roughly one thousand plantations around Berlín are in the process of closing up shop."

"Can they plant something else?" I asked.

"It is difficult," he said. "The costs involved in switching out of coffee are huge. Still, it may be our only choice. Oranges, cotton, who knows? There may be no market for them either, but at least we'd be diversified.

"Anyway," he said, "whatever we plant, we have a bigger problem. There is no one left here to work the fields. You can't even find two or three young people for a day's work. Everyone's gone."

"Could you point me toward any plantation still operating?" I asked. He just laughed and walked me to my car. He waved an arm through the air. When it reached north it stopped like the arrow of a compass.

"You won't find anybody," he said. "But if you do, tell them to come see me. I need them more than you do."

Despite Durán's pessimism I was determined to find some-body, anybody, who could explain why they were leaving, how they might leave and when. Whether or not they be-lieved in their hearts that they really could reach somewhere better. I needed that emblematic figure to anchor my story and set the narrative on its course.

I wound my way through the hills with their short, neatly planted trees drooping with beans. An old truck passed me and I thought to wave it down, but the man driving wore such a scowl that I balked. Later I passed by a ruined estate with a tall green gate behind which I thought I'd glimpsed a dog. I stopped and got out of the car. From behind the iron bars the animal barked at me ferociously, but I knew she had no serious plans because the gate itself was propped wide open. I walked in.

"Somebody must be feeding you," I said, kneeling. She had recently had a litter; her underbelly was tender and swollen and her teats were distended. I headed down the long grassy entry, uncut, past sheds and a large house that looked abandoned. The gray bitch followed at my heel, tail wagging.

There, across a sweeping yard shaded with decora-tive trees long unattended, I could make out a man and a woman. They were on the patio of a much smaller house—

a shack, really—square and squat and white. I approached them, waving.

"*Muy buenas tardes,*" I said.

"*Buenas tardes, señor,*" replied the man. He was youngish, maybe thirty.

"Do you live here?"

"We do, yes sir."

"I'm a reporter. From the U.S. I wanted to talk to some coffee growers."

The man's wife approached him, placing a hand on his shoulder. Four children were playing on the tiled patio behind them. The youngest one couldn't have been more than a year old.

"They're all gone," the man said. He said his name was Rafael Castellano. He wore a baggy blue button-down shirt and rough-cut leather sandals the same color as his mustache.

"They've all left," he repeated.

"For the United States."

"Where else does one go?" he said, smiling. His wife disappeared inside. When she came back out she was gripping a machete.

"Honduras?" I said.

"Honduras?"

"I don't know. Guatemala? Mexico? The U.S. is so far. Is it really worth it?"

"People leave for a better life, sir."

"Of course. I know that. And you? Are you leaving?"

"I cannot."

"Why not?"

"I am the foreman."

"The foreman. Of?"

"Of this plantation. Of all that you see here. The owner does not live here. He needs someone to organize the shifts, to oversee the harvest. We have over three hundred acres planted. It is my job to ensure that the cooperative pays us correctly. And to pay out the wages to the workers."

"But there is no one here."

"No. Just me and my family."

"You've got no one to pay. No one to supervise."

"No sir. I haven't had a single employee for some time."

"You might consider doing what your workers did," I suggested. "No one would blame you for abandoning your post."

"I don't have the money to leave," he said. "It is very expensive. And there are six of us. With my wage I barely can buy enough for us to eat. Now we scavenge for firewood to sell," he said, gesturing at his wife and her long cutting tool. "We are trapped here."

I looked around at the hills. At that bounty that had brought riches then ruin upon them. I studied the disrepair of the faraway owner's mansion.

"One day your boss might just give up before you do," I said.

"That day is coming."

"And then what?"

"Then we are finished," he said, shrugging. "Would you like some coffee?"

"I'm afraid I don't have anywhere to put it," I said. I

thought he was offering me a sack of the stuff. But he'd meant just a cup.

A couple of hours later I left Castellano's patio, jittery from too much caffeine, bidding farewell to him and his wife and their four happy children oblivious to their misery. And to the dog who, for the moment, still had this place to eat. I drove back into Berlin and then past it to a town called San Miguel and parked along the main, paved road. I would try one last time to find some would-be migrants before setting north myself.

San Miguel sat on a slight rise and the sun was setting. Those who left here needed only to follow the evening light toward the capital, about 120 miles away. That was a romantic enough notion, I thought. Perhaps enough so to boost a traveler's initial spirits. Not that this was by any means the most difficult part of the journey to the United States. This was, after all, still home turf.

I got out of my car. That slight hill rolling away from me, its slowly descending grade—a kid with a skateboard would have found the paved arc of it irresistible. I was having trouble myself. I was suddenly gripped by a childish urge to run down it. To catch up with the departed legions, to cross with them into something else. This trip, I wanted to believe, must deliver some reward. For the undertaking alone. Then up the hill there came a man walking. From a distance you could see something was wrong with him; he shambled slowly, listing to one side as if either wounded or very old.

"Excuse me, sir," I called from the curb as he passed. He wore a wide-brimmed palm hat and he carried a walking stick in one hand and a mango in the other. His nose was arced like a warlock's from a kid's book. He stopped before me and leaned his stick against his hip and immediately began peeling the mango, as if he'd just been waiting for an excuse to stop and eat.

"Where are you coming from?" I asked.

"From that way," he said, without gesture.

"Working the coffee?"

"There was no harvest this year," he said. "I am José Angel Hernández."

I introduced myself.

"We don't know why," he went on. "Perhaps God didn't want one." He was still talking about the harvest. He stood a full five strides from me but his shadow reached all the way to my feet. I did not ask him what I'd asked Castellano earlier. But he must have sensed the question.

"They say there are many rich cities up north, " he said, "But I am seventy-seven years old. It is too late for me to go."

"I imagine that you have family up north, though."

"I have grandkids, it is true. They have gone."

"So then you are okay. They are looking after you. Sending money."

"They don't send me anything. Young man, this life is one long struggle through shit. And then you grow old and you are forgotten."

"Aren't there any coffee workers left?"

"Heh, heh, heh," Hernandez laughed. He was chewing

crudely, moving his jaw up and down like a skeleton. Dark orange mango fibers were crisscrossed on his front teeth. "What do you think?" he asked. "Even the owners are gone! Nobody wants to be the last one to die here."

That idea caused me no small unease. Dying alone in your neglected house. I thought of my landlord Christy's mother back on Jojutla Street, sick in her bed before I'd arrived, then dying. Then I remembered how I'd fallen asleep on Christmas Eve in that same room, my ears straining for any last sound. A key, perhaps, fumbling in the downstairs lock.

The old man was watching me. "Me, I always stayed behind," he said without prompting. "While the others left, one by one."

"Why?"

"Because I was too poor. Or at least I thought so. And then one ordinary day it was too late. I'd grown old."

He said it like he'd been ambushed. My hopes now of interviewing anyone young enough to leave town without a cane were all but dashed. The youth of San Miguel had long raced off to save themselves from the immediacy of hunger, I thought. But also from tomorrow's oblivion, manifest before me now under this palm leaf hat.

"But it's not too late for you," Hernández said, to my surprise. "You're still young."

"For me?" I began. "I'm not looking . . ."

"¡Así que vete!" he barked at me loudly. *Scram!*

The violence of his outburst clipped my thoughts in two. When he stomped a foot toward me as if to shoo a dog I did start like one. He was smiling. I thought, José Angel Hernández might very well be demented. Certainly his tank of pluck had not yet run empty despite his lot. His mango was now finished, and he dropped the skin and the hard, slimy pit to the road and partly onto his feet. With a satisfied nod he shuffled off toward town.

I waited until he was out of sight then sat down on the curb. But not for long. I was still riding the slingshot. I got back in my car and took Hernández's advice. I struck out for the border between Guatemala and southern Mexico, about 350 miles away. Maybe along the way I'd get lucky and pass a hitchhiker, some young man or woman who might ask me if by chance I wasn't headed as far as San Diego or Dallas.

Either way I'd pick up the story again along the banks of a wide jungle river, home to no one but crossed by everyone heading to the States—or by those preying on them. There, there were dangers more concrete and immediate than the slow fade of towns like San Miguel from the maps. And surprises. Because the migrants from Central or South America figured the greatest peril awaited them at the U.S. border. This is what they planned for—psychologically, financially—but they were mistaken. Mexico itself was their biggest obstacle, their more determined enemy. Its thieves and thugs lay in wait on the banks of the Suchiate River, nearly reptilian in their pitilessness, for such herds to cross over.

Migrants do share this in common with all animals moving across great distances: at some point on their journey they are confronted by troublesome geography. This can include waterways that break the collective stride. Man and animal congregate before such banks, but sooner or later they know they must try to cross. Their other choices are to wither on the spot or be hunted down. Thus the choice is already made for them.

For Central and South Americans heading to the United States, one such waterway is the Suchiate, a lazy brown spill separating western Guatemala from southern Mexico. Its banks are all muck and mangrove and contraband, human or otherwise. When I first visited that place I knew that never in human history had a single written norm, regulation, or law ever been enforced there. There was simply no possibility of it. Therefore unwritten rules were followed closely, a group of migrants told me not long after leaving

San Miguel. I'd found them at last. We were sitting on the outdoor terrace of a migrants' shelter about a quarter mile from the river, on the Guatemalan side, resting in the shade on a hot afternoon.

Among the unwritten rules was never to approach the river at night. Even if the hollow whistle awoke you from sleep, you were not to heed it. Resisting that call was harder than it seemed. It tuned at intervals from across the water, and its ferrous echo carried with it all the possibilities of good fortune hoarded inside each migrant's imagination. But nothing was certain. The freight train's whistle might also be a harbinger of death. And because no migrant ever did know how his cards would be dealt, the whistle was disquieting and made returning to sleep difficult.

"The reason you do not strike out for the river at night," one migrant from Honduras was telling me, "is that you would not reach it with any of your provisions or the savings you'd set out with. You would be fleeced of such items and likely even your pants."

"I've lost pants before," I said.

"These things are much more valuable to you than to the crooks," he went on. "Food, your hidden cash, changes of clothes, identification, addresses of safe houses and relatives, photos to remind you why you are even doing this—because they will literally keep you alive during the two-week journey across Mexico."

The shelter I was being schooled in was called simply the Migrants' House. It was one of many in the area, set up by church groups and volunteers who gave talks to dozens of transients daily about their rights in Mexico, where to turn for help if necessary, what supplies they would need, how long the journey would take, and so on. But its principal mission was more basic: to let the northbound rest. Because if the migrants succeeded in climbing onto the freight trains passing through the yards outside the Mexican city of Tapachula, then their most immediate danger would be fatigue.

One man, named Rodolfo, had ridden the trains before. His life was in Los Angeles, California. He lived there with his wife and two children and said he drove a truck for a living. He had come back now to San Salvador because he couldn't bear not seeing his mother and father. He'd last seen them in 1989.

"The trains," he told me, "have boxcars that are not easy to ride on. Some are round. You cling to ladders on the sides, or sit on the roof or balance in the space between them. For hours and hours, deep into Mexican territory, hopefully past the army and police at the Isthmus of Tehuantepec. Then you are free.

"But you have to stay awake," he said, lounging on a wooden bench in the artificial breeze of a wooden ceiling fan. "If you fall asleep for even half a second you can fall off the train. And often this leads to your disfigurement or death.

"The only time you get to really rest," he said, "is when the train stops to hitch or unhitch boxcars. The best way to travel is in groups, remember that. With everyone keeping each other awake."

"*Oye*, but all that's *after* you've crossed the river," interjected another man seated on an old backpack. Like the rest, he was busy mopping up a plate of beans with a folded corn tortilla. "Julio César Suni, at your service. I have been back and forth several times from my native Honduras to the United States. But this part of the trip is when I am most nervous."

"Not because of the trains," I said.

"No, the river. First the river, then the trek to the train yard. I feel like a soccer ball every time I cross into Mexico. Because just about everybody tries to kick me. You're worth nothing over there. Even the Mexican police tell you that."

"The last time I tried to cross," Rodolfo interrupted, "they took everything I had. I will never forget."

"When was that?"

"Two weeks ago! I've been here since, resting. Waiting for money to arrive from my wife in California. I'd actually made it to the Mexican side. But then I hired one of those bicycle taxis to take me a couple of miles up to the road, to the train tracks. Through the jungle. Four guys with machetes surrounded me. I had no way of running. They took all of my cash. They wanted my sneakers too but I told them I didn't have anything else to wear. Somehow they let me go with my shoes still on. I thought I was finished."

As I chatted with Rodolfo and Julio César, others were slowly making their way over. Each man possessed a similar story. And the women's were all the more tragic, because too often a female migrant robbed in Tapachula found no other way to survive than by opening her legs for money—money to finance her return home or onward north, usually into the unsuspecting arms of her anxious husband. But on the terrace that afternoon the few women present did not speak of this possibility or, for that matter, at all. They sat straight-faced and silent.

Then from one moment to the next that lazing crowd sprang to its collective feet as if someone had run an electric current through the floor.

"What's happening?" I asked Rodolfo.

"It's time to go," he said. "The train's arriving."

"I didn't hear anything."

"Listen."

The men were gathering their knapsacks and jackets and some were hustling in and out of the bathroom, one last civilized ritual before suspending such pretense. I asked the director of the Migrants' House what was going on.

"The train is arriving," he told me.

"But I didn't hear anything."

"Listen."

"I *have* been listening!" I had my deck rolling still, hoping to catch at least one whistle as I followed the crowd down the outdoor stairs and into the central courtyard. Each shook hands with the waiting staff and filed out onto the dirt street. Another hard right and we were descending a muddy path toward the river's edge. The air was sweet and hot and the reeds grown up around us blocked the view toward the water. But you could hear voices. And the inanimate knocks and clangs of some brisk commerce. Then we were on the rocks above the muddy bank.

The first raft I saw, one among dozens, was crossing back from Mexico toward us. Like the rest, it was made from wooden planks lashed to several inflated truck-tire inner tubes. It was piled six feet high with plastic packs of toilet paper. Another heading our way carried cases of Coca-Cola and Fanta orange soda. Still another had goats on one side and a refrigerator on the other, in perfect balance.

A sole pilot poled each boat or pulled it by a rope through the shallower waters. All around us men were unloading such commodities as were needed in Guatemala while other men ordered us aboard the empty northbound vessels. Canned goods and cookies heading south, human beings heading north. The ferry operators and the middlemen must have been making fortunes, I thought. Some had guns visible in their waistlines. Wads of money were fanning everywhere, changing hands.

I climbed aboard one raft with Rodolfo. Our jeans were wet from the thighs down where we'd had to wade out a bit. We each handed the pilot the equivalent of about fifty cents. When he had eight passengers on his tilt-prone platform he scrambled aboard too and pushed us out into the river with his long pole. The current was barely perceptible, the water like heavily creamed coffee. Halfway across the river we had a better view of the international bridge spanning this chasm, officially connecting these two nations. It was right there, a quarter mile away, green in color and the only straight line in this jungle. Upon it, serious-looking customs officials from both countries studied passports and searched cars for illegal goods and drugs. As if such gestures were to be believed, given what was transpiring here below. What happens on the bridge, I thought, is little more than high theater.

"Don't *they* ever come down here?" I asked loudly enough for Rodolfo and the pilot to hear me.

"*Híjole,*" said our pilot. He was a chubby, middle-aged man with a horizontal scar on his forehead like a permanent worry. "They know better than that. They only come down here to collect their *mordidas.*" Bribes.

"I thought President Fox had a plan to tighten security on this border."

"Who are you?" the pilot asked. "Did the president send you?"

"Ha, ha, ha. No, no, no. I'm a radio reporter. I thought this border had been closed. At the request of the Americans."

"*Hijos de la chingada,*" he said. *Sons of bitches.* "Not even the entire *gringo* army could stop us. Do you know why?"

"Because they're fighting elsewhere? Because you've mined the jungle for miles around?"

"Because we're responding to a need," he said. "The *gringos* only want open markets when they control them. But they cannot control this one. If Mexico and Guatemala cannot control it, then no one can."

"Do you get a commission on the goods you bring back into Guatemala?"

"I'll tell you something, *Señor Gringo*."

"I'm listening."

"This river is a peaceful place."

"Yes. It's beautiful here."

"Do you know why you only hear the water and birds and not unpleasant sounds such as gunshots?"

"Would that be because nobody is bothering anybody?"

"*Órale.* That's right.

"It is not my intention to bother you."

The pole man set his pole down then and stepped toward me. He walked past, to the prow of the raft, and leapt into the water. It came up to his waist. He guided himself around the raft with one hand and grabbed hold of a waterlogged rope line and pulled us the remaining fifty feet to the shore.

One by one we jumped into the shallow water. I did not address our pilot again, nor did he speak to me. Instead I asked Rodolfo what his next move would be.

"To the tracks," he said. "Would you mind accompanying us?"

We walked together along a trail busy with migrants and merchants. Then there were fewer people. I wasn't altogether sure my presence would be any deterrent against miscreants. I was thinking maybe I'd only serve as chum. But twenty minutes later we had reached the rails without incident.

"So now what?" I said nervously. "To the stationhouse?"

"For you, yes. The police will not bother you. You will be safer there. Good luck." Rodolfo and Julio César and five others shook my hand and disappeared into the jungle. The sound of their footsteps stopped somewhere not too far off. I imagined them all just sitting down to wait. We were about half a mile to the north of town. The train that everyone but me had heard earlier in the day was apparently idling in the train yard back there, awaiting some loading or reconfiguration. It was midafternoon and even the sun seemed to hang still overhead. I picked my way along the railroad ties toward town, feeling like the only creature moving on earth. I thought of Jason Sprinkle, the artist, smacked from this life on tracks a world away from these. I did not cross a single person's path nor see or hear a train as I walked. When I reached the stationhouse I found it empty and locked.

So I sat down to wait like everyone else.

What we have been reduced to, I was thinking, half asleep. Men hiding like wild boars all throughout the day. Then at sundown, several hundred yards down the tracks, I saw a train approaching and many tiny figures leaping onto it. I leapt up myself and tore off down the tracks.

The train was moving very slowly still and I hoped I could jog alongside it and talk to the people interlocked on its ladders or peering from the open boxcars. But when I reached the locomotive I noticed something curious. The people leaping onto it were leaping right back off again. They must be training, I thought. But for what? This was it. The train was heading north.

I saw then that this train was very short, only an engine

and one car. It was not the full train but a section being moved from one place to another. I also noticed that all of the young men playing on it had something in common. None wore worried looks. They were laughing, covered in tattoos and some shirtless. And before I realized what I'd stumbled into, they had surrounded me.

Knowing the answer but unable to think of something better to say, I asked, "*¿Ustedes son migrantes?*" *Are you guys migrants?*

Two or three were right up in my face. The others circled around or threw rocks after the crawling train cars. I knew they would laugh before they laughed.

"Are *you* a migrant?" one asked me.

"I'm here to talk to them," I said.

"To talk to the migrants!" someone yelled. More laughing.

"Where are you from?" asked one guy wearing a New York Yankees cap.

"Why are we talking to this guy?" said another.

"Let's take his shit. Look at those cameras."

"Let's fuck him up."

"Let's beat his ass."

"We could open him with a knife."

"What is that anyway?"

"It's a TV camera!"

"No it isn't," I said. "It's for recording sound. No images."

"But the cameras."

"Yes, these are for images. But I haven't taken any."

"Where are you from?" repeated the young man in the

Yankees hat. He stood to the back of the group. He was bigger and a bit older.

"New York," I said.

"Yeah?" the guy said, switching to English. "I lived in New York for three years."

"Yeah? Where?"

"The Bronx."

"*Chévere.* I'm from Pelham."

"I know it. Pelham Bay."

"I wanted to interview immigrants," I said.

"None of us are migrating anywhere," he said. "So I guess you're out of luck."

"You live here?"

"We do business here," he said.

"Where are you from originally?"

"San Salvador. But my home is called La Mara Salvatrucha."

I nodded. The Salvatruchas was one of Central America's largest and most vicious street gangs. It had hundreds of thousands of members in chapters reaching all the way into the United States and Canada.

"And your business here?" I asked.

"We look for opportunities as they arise."

"Why are we talking to this guy?"

"*Oye,* let's take his things!"

"Nobody is going to touch my boy," the gang member said, switching back to Spanish. He approached me and said, "Listen, I'm going to walk you back to the station. Don't listen to them, and don't even think about wising back."

We set off, just the two of us.

"If you don't mind my asking, do you shake down the immigrants?"

"Do you go fishing where you know there are no fish?"

"Huh?"

"There are no saints left in this world," he said. "We cannot really discuss these things together and hope to find understanding. You journalists are always looking for stories of redemption. Our road leads elsewhere, and we are not open to miracles. So go, there's the station. I recommend staying on the platform. Soon the police will arrive. They usually come at night."

"Thank you for not assaulting me," I said.

"You're welcome. Do not come back."

It was fully dark now, and I sat down to settle my heart rate some. The *maras* of Central America did not normally show such cordiality toward strangers. I was thinking about what he'd said. No miracles here, no redemption. I leaned back on my forearms and heard the sound of a match strike.

Sitting at the far end of the platform was a figure, a man was all I could tell. In keeping with the day's theme I decided to approach him. Slowly. Making some noise as I did so.

I sat down a short distance away and said, *"Hola."*

The man, in a white tank-top undershirt, his shoulders slouched and his head shaved, turned his head slowly, already smiling. He smiled slyly as if he knew what I wanted, knew why I was here. Surely I was not the first reporter. In his fingers he held a half-smoked joint. He held it out to me.

"Wanna puff?"

"*¿Porqué no?*"

I slid over next to him and slipped the flimsy joint from between his fingers. The back of his left hand was a swirl of scars and what appeared to be early attempts at tattooing. He was muscled and lean, and the black stubble of his head was interrupted by other scars where no hair grew. I wondered what he carried in the pockets of his baggy camouflage pants.

"*¿Usted pertenece a una pandilla?*" I asked. *Are you a member of a gang?*

"Yes, I am," he said. "Do you know the Calle Diez y Ocho? It is a gang. I will admit this to anyone who asks. I do not apologize for it. If you watch these parts you will see all manner of unpleasantness. All manner of people." He took back his joint. "Here anything and everything happens, do you understand? All kinds of gangs are moving through here. Just as they do throughout the world. We are everywhere. And in each place we manifest a different face. But here our presence is especially strong."

I nodded.

"Do you read the Bible?"

"I have," I said, exhaling a ball of sweet smoke.

"In the Good Book it says that the word *mara* comes from *amargura*. Bitterness. The Calle Diez y Ocho has taken us in. Or more correctly, it has taken us. We have come up from the south while you have arrived from the north. The bitterness is found exactly here, in this place where we meet.

There have been many witnesses. We assault and in turn we are assaulted, back and forth. Some among us would even attack a stranger like you in order to raise our voices to whoever listens and proclaim 'I am a thug, I am a killer.' But if you say this out loud you're lying to cover up what you're not. We in the Calle Diez y Ocho, we move in silence, proclaiming nothing. Which is why those riding the trains hold us in such high regard. We appear like leaping phantoms, armed with supernatural powers of persuasion."

He laughed. This was serious stoned talk. Then I heard for the first time the sound I'd been waiting for: the piping, inveterate whistle of the train. I looked down the tracks and was surprised to see the engine grinding slowly toward us not a hundred yards away, nearly imperceptible. From the forest all around, on either side, hundreds of shadows were emerging. Dark figures that hesitated, then began to walk or jog alongside the locomotive.

"Do you understand me?" asked my interlocutor. "We are not heading north. It is you who has come to us. Why is that? This train is not yours for the riding. She is a beast. Do you know what it means to be truly challenged? All of the *illegales* you see before us, they defy the beast because they need her. As do we. But this train is a people-killer. Make no mistake. She has free rein and no one can stop her but God. Remember what I'm telling you."

He stood and licked his fingers and pinched out his joint, stuffing it into one of the smaller pockets on his pants. The train was in front of us now and the man leapt casually to the gravel before it. Like the rest, he began jogging along, looking for a ladder or window to jump to, blending in. I climbed down and began jogging too. The train's shell was filling quickly now with people. It looked like some huge segmented insect under siege by ants. Hundreds of hunters bringing down their prey. But of course the truth was the other way. These tiny, silent riders were the hunted. Threatened by man and machine alike. Fatigue might cause them to nod off and fall to the wheels below. Or the *maras* might get them, which implied beatings and theft and in the heat of such criminal enterprise perhaps even a free toss to the ground. And though the police had not appeared so far this night, they still might, down the tracks. They too were accustomed to taking their fill, the only difference being their badges.

The train was gaining speed now as it reentered the jungle beyond the stationhouse. I slowed to a walk, awed, slightly stoned, watching it and the multitudes upon it disappear into the black foliage. Before the jungle had entirely enveloped the boxcars I saw some people jumping off. Migrants who'd lost their nerve or gang members who'd already struck, it was impossible to know. This time I did not run forward with questions. I'd received more than my fair share of clemency for one day.

The next morning at another migrant shelter, here now on the Mexican side, I found convalescing those who had not found merciful passage on the Tapachula freighters. Some had arrived here discharged from Tapachula General Hospital with bullet wounds or deep machete gashes. But most of the two dozen Central Americans at the Good Shepherd Home had had their journeys cut short because the trains had literally done the same to them.

Alma Belinda Cruz, a heavy-set thirty-year-old mother of two from Tegucigalpa, Honduras, sat at a table with copies of newspaper articles spread out before her. She was nearly healed, nearly ready to return home. It was raining, and buckets had been placed about the main hall to capture what water dripped in.

"This is what they wrote about me," Cruz said proudly. There was her name, I could see, in a headline, in the first graph of an article, beneath photos. She leafed through the copies, holding each one up for me to examine. There was a photo of her smiling, another of her laid out on a bed. One showed her two amputated legs, severed above the knee, lying on what appeared to be a medical gurney.

"I was working in a fast-food chicken restaurant in Honduras," she told me. "But then the layoffs came and I had no way to support my daughters. So I decided to come north, alone. To make money and send it home."

A nurse came over and gave Alma a small paper cup with pills in it. She swallowed them with water.

"My friends and I were hiding in the jungle about an hour and a half north of Tapachula. It was February 4 of this year. Nobody back home had told me we were going to jump a train. But when we reached Mexico one guy in our group said the train would be faster. Thus it was decided. I was unsure, but he said my other choice was to turn back. All right, I said. I'll give it a try. I'll see if I can do this. I said, God is going to protect me. He is going to help me because I have been a good person. I wanted to find a job, to change our economic situation for the better. For the good of my mother and my daughters. But it didn't work out that way.

"When the train came past we ran out of the jungle. My companions all jumped aboard. There were twelve of us, and we knew each other from Honduras but not overly. I was running along next to a ladder and lunged for it. I grabbed it with my hands, but when I jumped my feet missed the rung and my legs swung full under the train. I am a bit over-weight. I lost my grip and fell.

"The train never stopped and my companions did not jump down to help me. I think perhaps they didn't notice because I didn't scream or say anything. But luckily there were other migrants. Three boys came and gave me first aid as they could and took me to a hospital. I never once lost consciousness. This leg, the shorter one, the train cut it clean through. But the other one was still attached. The boys who helped me, they put my leg in a knapsack and carried it and me together. I understood everything they were saying. I could hear the doctors, too.

"That is how I know that I am extremely lucky to be alive. To be alive and to be able to go back home and see my daughters again. To take care of them."

A young man was wheeled over to the table and positioned next to Cruz. He began eating soup. His right arm
from the elbow and his right leg from the knee, gone. There
was a man in a wheelchair nearby, his back fractured, reading a book. Down a short hallway, behind other closed
doors, the less-healed.

Alma Belinda Cruz spoke of taking care of her daughters,
but I wasn't sure how. She'd found that task impossible *with*
legs. So out of desperation she had done the unthinkable.
She had walked away from her children to save them. I wondered what on earth she could do for a job now, confined to
a chair. What any of these damaged people could do once
they got home. If the whole-bodied couldn't make a go of
it, you had to assume such amputees would be doomed to
penury far worse than what had driven them from home in
the first place.

I said good-bye and good health and left that rare nook
of angels, traveling north again to catch up with the vast
stream of migrants who'd held to the trains or skirted them
entirely.

In a market of makeshift stalls in the desert town of Altar, in Sonora, Mexico, a slight-of-build Oaxacan farmer was shopping for pants under a sun sweating us like raisins.

"What is your size, *señor*?"

"31, *señor*."

"*Bueno*. We have much to choose from in size 31."

"I don't want size 31, *señor*."

"What? Oh, I see. You are shopping for someone else."

"No, they are for me. Please, give me a size 36 jeans."

"36?"

"36."

"But a size 36 won't stay up!"

"Don't your jeans have belt loops?"

"Do not get in a huff, *señor*. I have size 36. I have size 42 if you'd like. I also sell belts."

The vendor put a pair of 36-waist jeans into a green plastic bag and handed it to the farmer, who stuffed it into his pack and then paid. On the outside of his pack two gallon-sized plastic jugs were tied to loops. At the moment they were empty.

"Excuse me," I said quietly to the man after he had walked out of earshot of the stall. The main square in Altar was not a place where you particularly wanted to stand out. "May I have a brief word with you?" I asked. "I am a reporter for the radio. I don't care about your name if you don't care to give it. I just want to know why you bought pants that won't fit you."

"But they will fit me," he said, tugging at the jeans he was already wearing. "Over these. When you cross the desert you must wear two pairs of pants. For the nights. Against the cold and the scorpion stingers."

The farmer said this was to be his first time crossing the desert. But he'd gleaned the pants tip and many others from friends who'd crossed before and from those he'd met on his way north from the Pacific Coast. That network of knowl-edge sharing is what had landed him in Altar as well. This was the surest place to arrange your crossing, the most tried-and-true launching pad.

Altar was one vast smugglers' bazaar, swarming with U.S.-bound immigrants and the traffickers who knew the ways in, all of them trying to connect with each other under the constant fear of trickery or official observation. With-out this multimillion-dollar commerce, Altar would have been but a dusty, one-lane outpost. There would have been no constant traffic of buses and taxis arriving from airports to the south, none of the attendant cafés and taco stands, no stores hawking survival ware, no paved roads or houses rising on the sand around. Illegal border crossings had put Altar on the map, and now it drew hundreds of thousands of immigrants each year.

But we were still an hour's drive from the border. To cover that final leg of the journey, drivers were packing immigrants into dozens of minivans on the streets surrounding the square. As the vans pulled out others were arriving empty, having dropped their loads in strategic border towns such as Naco and Sasabe. They moved back and forth all day long, like military convoys moving troops to the front.

In my own car, I followed a line of vans heading toward Sasabe. I had a meeting set up later in the day with the U.S. Border Patrol in Tucson, Arizona, and Sasabe was more or less on the way. I stopped there for lunch in its unbearable brightness, the walls of houses, the sand, everything beige and glaring. The sky was a whitish blue that accosted your eyes. Hundreds of immigrants wandered the town's few hot streets, listening to local bands playing *ranchera* music or to the same music on radios. The least informed among them were drinking beer in their last hours before leaving.

I had not traveled by freight train across Mexico, but it was easy to read that journey's mark on the faces of the men and women around me. They were drawn, exhausted. Some still seemed to be in shock; some were smiling, not for joy's sake but because they were not maimed or awaiting deportation in some prison yard.

The trains that had delivered this latest wavelet of desperate travelers had rolled across the country day and night for nearly two weeks, with minimal repose, winding from Chiapas to Tonala, that southern city built on clay where the sun first emerges, according to the Nahuatl meaning. Then beyond to Mexico City, Guadalajara, Monterrey, Chihuahua, Hermosillo. From there some migrants continued farther north to Nogales or Juárez, sitting ducks in a desertland of injustice.

The migrants' final stopping point within Mexico was largely a matter of chance. They were carried along in accordance with the unpredictable track switches of engineers responding in turn to the vicissitudes of the distant market forces that controlled them. The immigrants held sway over none of this. They simply held on as best they could, then tried to descend from their transport as close to the border towns as possible, wherever that might be. And if when they finally did reach the border it was still daylight, they would settle back to wait once again. Just as they had near the banks of the Suchiate River, in Guatemala. To wait for the decisive night, the once-and-for-all night of this final crossing.

In Sasabe, the air seemed filled with excitement. With the border so close at last, everyone was anxious, and relieved to have put the possibility of death on a train behind them.

I could not wait around until nightfall. I needed to get to Tucson for my meeting with the Border Patrol. After lunch

I crossed the Mexico-U.S. border through the Sasabe port of entry. It was 100 degrees out. I rolled easily up the South Sasabe Road to Highway 289, then onto the Arivaca Road and merged with State Highway 19 for the final stretch into Tucson.

This protected swath of the Sonora Desert was carpeted thickly with prickly pear cacti, mesquite, catclaw, and saltbush. It sounds strange to call a desert a jungle, but if you've walked there you know. The sun was making a bright line of the road ahead, but a half mile to the east hung thick brown clouds and it was pouring rain. Lightning sketched across the dark sky.

To the west, out the driver's side window, I began to notice all manner of clothing caught up in the near brush. Some articles lay on the highway itself. As I drove I passed more and more items out of place on the asphalt. Shirts, a shoe, two old backpacks next to each other, their frames to the sky like upended turtles. I slammed on the brakes, pulled over to the shoulder, and grabbed my gear, pushing the record button even before opening the door, then stepping out into the heat. It looked like a small plane had crashed. The whole unnatural scar ran for a hundred yards. The vegetation was torn free all the way to the Chevy model van, which was upright now by pure chance but smashed, immovable, its windows gone, its doors gone except for the back ones hanging broken.

There was only one ambulance on the scene and some unlucky Samaritan or two trying their best to comfort or revive the plenitude of wounded. Bodies everywhere. The falsetto moaning of men. I ran across the street and knelt beside the first man I reached. I set my gear down and touched his shoulder lightly.

"*¿Llegamos a Los Angeles?*" he was whispering over and over. *Did we make LA? Did we make LA?*

"*Todavía no,*" I told him quietly. *Not yet.* It was hard to see just where the wound was on his thick head. He was also injured in the groin or upper leg. The coarse sand was not absorbing his spillage. "Try to lie still," I told him. I was going to pass my hands along his length feeling for broken bones and openings.

"Get the hell away from him!"

I nearly fell backward.

"You goddamned journalist piece of shit get away from my wounded!"

A tall man, American, with a beard and a blue windbreaker bearing the logo of some local agency, was pointing a long finger at me. In his other hand he held a crackling walkie-talkie.

"I was trying to help," I said. "I can help you if you need it."

"You can help me by getting to the other goddamned side of the fucking road. Now! Move!"

I stood up and grabbed my gear and ran past that lifesaving volunteer fireman son of a bitch all the way to the busted-up van. There was a woman seated now on the back bumper holding a bloody towel to her head.

I sat down next to her. So far I had counted eighteen

people. Two were not moving at all. One man, next to the van, was bent such that I was sure he would not stir no matter the medical prompting.

"*Señora*, do you feel well enough to talk?"

"*Sí.*"

"Can you tell me what happened?"

"Yes, I can. What I know of it. We were just driving along. I couldn't see anything. We were all lying on the floor, pressed together. Hiding. We were going along and then there was a bang and we swerved and then the van started rolling. The next thing I knew, I was lying on the ground, far from the vehicle."

"How many of you are there?"

"Twenty-three," she said. "Plus the driver."

As we spoke several police cars arrived, and another ambulance. I could hear a distant chopper. When it was just a couple of hundred yards away from us it pulled up a bit, like a horse abruptly reined, then hovered. The guy with the walkie-talkie was talking into it and making hand signals at the helicopter pilot.

"Twenty-three?" I said. "In this van? It's made for, what? Ten max?"

"There were twenty-three of us waiting at the meeting point in the desert when the van arrived. There was an argument between the driver and our guide, but then he agreed to take us all."

This woman, her name was Cristina, said she was from El Salvador. She and her group had walked three days in the desert to reach a designated pickup spot. From there they were supposed to make their way in this van to a safe house in Tucson, where they would remain in hiding for several more days before slowly filtering out.

"None of us in the van knew each other," she said. "I think one of the tires blew. I thank God that I am still alive."

"Where is the driver?" I asked.

"He left with the others, helping them carry the child."

"There was a child?"

"The wounded little girl. Two men did not appear to be hurt. They picked her up and ran into the high brush. The driver helped them. He said he knew where to go."

So that's what the helicopter was for. To track down those who'd limped off. To detain them. To help that girl, whatever her injuries. Now the sirens were warping the air and there was a sufficient number of men to care for all of the injured. One emergency worker unfolded white sheets and lay them over the bodies of three passengers. Clear IV bottles glinted in the sun, the storm still raking the desert in the near distance. Stretchers moved victims toward the ambulances as the first drops of rain pattered against us. And out of the desert, on the far side of the highway right next to my car, a second group of immigrants appeared.

They ran across the road waving their arms. There were six of them, all teenaged boys. Their lips were gray and cracked and they could barely talk for their swollen tongues. They asked for water, water, and more water. It took them a long time to slake their desperate thirst. When they could talk, they told Border Patrol agents how their guide had taken their money and abandoned them on their first night in the desert.

"We've spent the last two days hiding in whatever shade we could find," one said. "At night we climbed ridges looking for a road, for lights, for any signs of life."

"But it was the sirens," said another. His nose was caked with blackened mucus. "*Híjole*, I thought we were going to die out there."

It was then that these youngsters seemed to first notice their catastrophic surroundings. The van and the white sheets like an outdoor morgue. This mortal unfolding that had killed some but led them to safety. Not an element at this scene conforming to their idyllic imaginings of what a trip to the United States might hold.

"*Sólo quiero irme a casa,*" another boy said, sitting down cross-legged. He started to cry a little. *I just want to go home.*

So did I, truth be told. I was tired now after many weeks on the road, and I'd recorded enough sound. If I couldn't put together a three-part series to make a few public radio listeners sit up and forget our wars for just a moment, I'd quit the business.

That evening I drove back into Sasabe like a man returning from the future. In town the party was still going, the mood festive. I knew from statistics that most of these people would make it safely across and into the cities, but still I couldn't go along with the revelry. Not after the broiling desert and the violence that had dropped in. I walked to the edge of town in the dark and found a stone altar to the Virgin of Guadalupe at a fork in the road. A group of migrants was there, holding candles, praying out loud to their protectress, praying for guidance as they prepared to deliver themselves to their wily desert guides. When they finished they bowed their heads and traced invisible crosses from their foreheads to their chests and across both shoulders. Then they brought their crossed thumbs and forefingers to their lips to kiss and made the cross again. Now they were ready. They blew out their lights and moved off wordlessly toward the border, less than a mile away. I followed behind. Until one man turned around.

"Please do not follow us," he whispered. "*Por favor.* Down this way there are thieves and worse. We are only taking this path because we have no other choice. But you . . ." He did not finish his sentence.

I thought for a moment about what he'd said, then took stock of my circumstances. It was nearly pitch black and I had no flashlight for returning. So I stopped there and just watched them go, rolling that last lick of tape before I packed it in: the sandy scuffle of feet fading into the darkness. Hope's footsteps diminishing, but never the hope itself.

When I landed at the Mexico City airport I dumped myself into a cab like pasta into a colander. Every muscle in my body was limp with fatigue. I was thinking about the series but also about all the plants I would have to replace at the house. Azaleas and bougainvilleas on the terrace, spider plants and ficus inside, withered all. I'd been gone for weeks, and no one had been there to water them. Walter and Guadalupe and Karina were all living in Acapulco now. Lazaro Roque was far down some other path, and no other refugees had arrived to take his bed. The house stood as empty as ever, but I knew I could stand being home now—and stay home for longer than a quick shower and a change of clothes. Enough time had passed. That's what I felt in my gut. But when my taxi pulled up in front of Jojutla 13, and when I saw all the lights blazing again through the second-floor windows, my heart betrayed me as before.

My plants had not all died. In fact they were doing extraordinarily well—better, certainly, than under my haphazard stewardship. The upstairs terrace was abloom in orange- and lilac-colored bougainvillea and the bushes' green reach had extended far in both directions along the outer wall. It looked like spring even at summer's end. The azaleas were thick and leafy. Even my cacti seem to resonate with color and life.

"How did you manage it?" I asked.

"I put water on them."

"That's it? You didn't talk to them, sing to them, caress their leaves?"

"No."

"Because that's what I do."

"*Oui*, when I first came in, I discovered how well your method works."

"When did you first come in?"

"The first time, a couple of months ago. I thought I couldn't save them. I came by every two or three days after that. Then Guillaume discovered our emails."

"What?" I sat up. "How is that possible?" We were lying on a blanket on the roof. The same roof where my bottle rockets had fizzled so miserably last Christmas. I'd been home less than an hour, and so far I hadn't dared ask her what was going on. Whether she would be leaving again.

"You hadn't erased them?"

"I did. I put them in the computer's trash."

"And he went through the trash."

"He went through the trash."

"That's not erasing."

Anne didn't respond.

"Anyway our emails are from before. From before I went away. When I left we were finished. That's why I left. What could he say about them now?"

"He said many things. He picked up a big knife in the kitchen and stabbed it through a wheel of cheese. He said he was going to kill you. He called me a whore."

"He must think I've stolen you."

"He said that, exactly. He blames you. And he no longer trusts me. Two weeks ago he threw me out."

"*Por Dios*, you should have called me. Where did you go?"

"I got my own place."

My heart sank a little. "Oh."

"It's something I need to do," Anne said. "I need my own space. That we can see."

"I understand," I said. "But I'll have to talk to Guillaume.
I didn't steal you away. You can't just steal somebody, like a
horse."

I said this out loud more to convince myself than Anne.
Now that our truncated affair had apparently blown up in our
faces—that worst of all possible scenarios—I was mortified.
Had I really stolen her? Why hadn't I just waited for her to
leave him? It was true that we'd thrown in the towel but not
before some certain intimacy. And Guillaume had stumbled
upon the words of it. Now he was surely coming for me, knife
in hand. Or without a knife. It didn't matter. He had a far
stronger weapon against me, which was the high road.

Anne and I went to see a shaman south of Mexico City to
cleanse ourselves of the guilt haranguing us. We had each
hurt Guillaume in different ways, and that could hardly be
some commonality to bring us closer. I began to feel as if
Guillaume were accompanying us everywhere. In the car, in
the kitchen, in bed.

The shaman's name was José, and he had built on his
debris-strewn rural ranch a large, igloolike structure made
of stones and clay called a *temazcal*. It's a sweat lodge, but not
for a group; the *temazcal* is a solitary space, a solitary experi-
ence, even when the shaman is in there with you. As José's
wife was heating its stones with fire he took us aside sepa-
rately, first Anne, then me, leading us into a slotted wood-
shed. There we stripped. He rubbed our bodies vigorously

from the feet to the crown of the head with an uncracked egg. I thought mine might break under such stress, but it didn't—not until José opened it into a glass. He studied the viscous white, the cloudiness in it, for some sign about which he did not speak. Then he ordered Anne, slightly embarrassed in her panties, to crawl into the *temazcal*.

Steam came from the entrance and there was faint chanting from both of them and the sound of brush or branches rustling. Ten minutes later Anne came out and tried to hide a smile building at the corner of her mouth. She sat down on the floor, with a towel around her, looking at the ground. José nodded at me.

I could barely shimmy through the entrance. When José was inside the dark center of his dome with me, he told me to lie *boca arriba*, on my back, looking up. It was very hot and humid and he kept dousing searing rocks somewhere with more water. I tried to relax, but José had seized the branches I'd heard earlier and was raking them up and down the length of my body. Then he was pounding me with them. Laurel or eucalyptus. Punishing or purifying. He was humming slightly. Occasionally he would stop and clear his throat, then spit into what sounded, in my blindness, like a metal bucket. The throat clearing grew more frequent as the minutes passed. Then José laid himself next to me and held fast to my right arm just above the wrist.

"Chant with me," he croaked.

It was not really a chant. The sound we made was like a child imitating an accelerating car, the note starting low and rising to our highest respective levels. I was feeling a bit un-anchored, and soon José was being racked by what sounded like phlegm-soaked seizures of the throat. I continued to make racing car noises despite my companion's gargling, shifting back to low gear and starting again when my voice cracked. I had a horrible feeling like one of José's hot rocks in my stomach. Soon I could no longer distinguish between the heat inside and out. It rose up through my chest and into my throat as my chant rose. Then it reached my head and I temporarily lost most of my sense of hearing. I was just plain scared now. José was digging his fingers into my forearm and gagging on his bilious fluids. There was a distant and nearly continuous splatter of liquid into his invisible bucket.

I was stuck on a high note when my brain went squishy. Something had snuck inside my skull. The pressure build-ing, the sound disappearing. I felt reduced to that pinprick of light on an old tube television that's just been turned off. Jose was puking and hacking, and then to my great relief he was pushing me out of the *temazcal*. I rolled over and scrambled as best I could on my belly across that floor of dry leaves and sticks and sloughed into the light, sweating and naked. José slithered out and his wife handed him a wooden bucket filled with cold water. He splashed it across my red, inflamed back. Then he did it again. And again.

I tried to cry out. You bastard. Please. I'm sorry. I'm so sorry. But I just broke into tears. I sobbed and sobbed. Anne sat an arm's length away, watching me with concern. I could see myself from outside my body. Pale and diminished, one clothes-size smaller, my head on the ground and my hands over it. At last the racking subsided, and when it did I was surprised to feel the heaviness lift away. The heaviness of betrayal and of so much uncertainty, I supposed. José put a towel over me. But I didn't move. I didn't want to yet. My head was starting to clear, and washing over me now was the same relief I'd felt after that spirit sat on me on my bed, on the eve of my trip to Colombia, in September 2001. Before the world had changed.

But finally I did stand up. Because the world had changed, and in so many ways. My world. I thought of the men and women waiting to cross the Suchiate River in Guatemala, how they'd had no choice but to move forward. Now, neither did Anne or I. We had been confronted by our own troublesome geography, as it were, and today was meant to be our crossing over. I pulled myself up from the shaman José's damp patio and Anne and I got dressed and got in the car.

"What was with the egg?" she asked.

"I haven't the slightest," I said, grinning.

We rolled down the windows and drove home.

Some time later, as Anne and I were moving her things into the house on Jojutla Street the most unexpected news arrived from Washington, D.C. The Bush administration suddenly wanted to talk immigration again. It came across the wires one evening as I sat at my desk. More than two years after September 11 the Americans had apparently reached the conclusion that sealing the Mexican border was impossible and pointless as a means to secure the country from further terrorist attacks. It was simply too late. There were an estimated eight to ten million Latinos already living inside the United States illegally, about half of them Mexican. All of these migrants were flying below Homeland Security's radar.

Finally, someone suggested what President Vicente Fox had been arguing for all along: why not create an orderly guest-worker program that lets registered participants work in the United States and cross the border at will? A high-level Bush team drummed up some details, and Mr. Bush himself went on TV to talk about it. The two leaders agreed to meet at a summit in Monterrey to discuss further details. I hadn't seen this coming. Nobody had. I was thrilled.

"Anne! We're back on the radar!"

I ran down the stairs. I found her in the bathroom with a pill in one hand and a glass of water in the other.

"Hey," I said, "why don't you chuck those in the toilet?"

"Because if I do that, you'll have to wear a condom."

"Bush and Fox are friends again," I said.

"How wonderful."

"I know!"

"That's why you were yelling?"

I closed her hand around the birth control pill. "Seriously," I said. "What if you were to get pregnant?"

"Then we'd have a baby."

"I know. But would you say it like that? Or would you say it with a bit more enthusiasm?"

"I'd say it with a bit more enthusiasm," she smiled.

I opened her hand and guided her to the toilet and turned it over. The pill plinked into the water.

"Let's just see what happens," I said. "For all we know, I'm shooting blanks."

"What does that mean?"

"That I can't get you pregnant."

Six weeks later we bought a home pregnancy test kit at the local drug store and watched the bold red line appear like a life itself. A local doctor estimated the fetus was about one month old, too early really to share the news, but we did anyway. Then something went wrong. On a lazy Sunday morning Anne awoke with a disquieting amount of blood on herself.

She had no gynecologist in Mexico, so she picked up the phone, and after a series of calls to friends she learned of a pregnancy clinic open on Sundays. I helped Anne into our car and sped to the address. At the clinic I sat her in a chair. "Wait here."

I went to the reception area where five young female receptionists waited in their white uniforms.

"My partner is pregnant. One month. Today she started bleeding."

"Have a seat, please," one receptionist told me. She handed me a medical form to fill out. "The doctor will see you in just a moment."

Another accompanied me back to the waiting area to speak with Anne.

"You must drink a lot of fluids," she told her. "And do not urinate."

"Right back!" I ran out to buy a two-liter bottle of water at a nearby gas station. An hour and a half later the bottle was long empty. Anne was at the bursting point, her bladder ready to rupture.

"Listen," I told the women at reception again, "I am not kidding. We've got an emergency here. My partner is six weeks pregnant and she's bleeding in a way she shouldn't be."

"As soon as the doctor is free she'll see you."

"When might that be?" Anne asked. I hadn't heard her approach the reception area.

"*Ahorita.*" *Very soon.*

We sat back down again. There was a bathroom door right in front of us. "Don't look at it," I said. I tried not to look at my watch. Twenty minutes later I lost my nerve.

"I want to speak to the doctor right now," I said.

"She is unavailable, sir."

"Make her available then. It's been two hours. If she knew there was a patient here with urgent complications I'm sure she'd attend to us."

"I cannot do that."

"Then tell me what room she's in."

"She is not here."

"What?"

"She is not here."

"Where is she?"

"*No sabría decirle.*" *I would not know what to tell you.*

"Is she in Mexico?" My shouting was attracting some attention.

"She should be on her way in to the clinic, sir."

"Has she been here at all today?"

"Not yet."

"And out of the five of you, none thought to tell us this?!"

Anne was now at my side, her eyes nearly crossed by her need to pee.

"*¿Qué onda?*" she asked. *What's going on?*

"There's no doctor at this clinic."

"What sort of incompetence is this?!" she exploded. "Is this a joke? I am bleeding! I may be losing my baby!"

"*Siéntate, mi princesita,*" one receptionist said flatly. *Take a seat, my little princess.*

"Little princess?!" Anne repeated. "Have you no conscience?"

"There are other people waiting here too, *princesita.*"

"Are they pregnant and bleeding from the vagina?! Do you have any sense of what an emergency is?"

"*Cálmate, Anne,*" I said. I was worried if she kept screaming she was going to complicate our complications, medically but even otherwise.

"The doctor will see you when she arrives," repeated another receptionist, stepping forward.

"Listen," I said. "Where's the nearest hospital emergency room?"

"I would not know what to tell you."

"You don't know?"

"I would not know what to tell you."

The rest just shrugged in unison, smiling slightly. Smiles of ingrained, inscrutable politeness designed to deflect our fury and desperation, I thought.

Anne headed straight for the bathroom.

We were back in the car, Anne was on the phone, I was driving to the house for lack of a better plan.

"Turn here," Anne said suddenly. She was getting directions to a local hospital from another girlfriend. We had wanted to avoid the city's notorious emergency rooms, but now we had little choice.

"Where are we going?"

"To see a doctor. Here it is! Stop, stop."

Twenty minutes later a pediatrician was leading Anne into an x-ray chamber.

"What's that?" I asked when the image was ready. You could see two white lines leading from the vicinity of the fetus across the uterus.

"That is the bleeding," the doctor said.

"Is it stopping?" Anne asked.

"Not yet. This is touch and go. I am not sure whether you will keep this one." The doctor said it kindly, as if there would be so many to choose from. "What you must do is go home and lie in bed. You must lie down for a week."

"And when will we know if everything is okay?"

"After about a week. I'll give you an appointment for a week from tomorrow, Monday. The important thing is to rest and relax and stay horizontal. Your husband will cook for you and help you. You should only get up to use the bathroom."

I walked Anne to the car. She was distraught. The whole way home she stared out the window, occasionally wiping tears from her cheeks. She refused my hand, then took it and held on to it tightly. At the house I spruced up the couch a bit and placed the TV controls within arm's reach.

"What would you like for lunch?" I was asking when the office phone rang. I bolted up the stairs.

"Gerry, it's Didi."

"What's up?"

"We need you back in Haiti, ASAP."

.

While Anne was bleeding, Haiti had gone into cardiac arrest. For over a month now the capital had been inundated with near-daily protests. A hundred thousand people for Aristide, a hundred thousand against; five hundred thousand for, five hundred thousand against: each march masked by festive music until the violence. The temperature was rising in that sealed cauldron, and people were starting to seriously question whether Aristide could in fact ride this crisis out and hang on to power. With each day the possibility looked more remote. The seeds of disaster sowed during the fraudulent legislative elections in May 2000 had at last broken soil.

People were dying in those daily street clashes. The country's opposition leaders, far from accepting the president's belated offer to hold new legislative elections, were out front and leading huge antigovernment marches. Clearly they felt that the tide was moving this catastrophe in their favor. As loved as Aristide was by many Haitians, even he couldn't govern ad infinitum in a country where every institution was penniless and the economy had ground to a standstill. His long-running campaign slogan had been "Peace in the mind, peace in the belly." Since his election he'd been unable—or, as some argued, not allowed—to deliver on either promise.

As the pressure built, the international community sty-
mied any chance for political compromise by refusing to
force Aristide's opponents to the negotiating table. This
would have been as simple as a nod from President Bush.
Haiti's opposition politicians were, for the most part, un-
knowns who would have had no leverage at all against Aristide if the United States and others had not discovered in
them a most convenient pawn.

But Bush's gaze was on other calamities far more press-
ing. Thus, tiny, insignificant Haiti—that fantastically castigated half-land, its offspring of slaves forever bequeathed
and bequeathing tragedy, squeezed as always by the more
powerful—erupted in bloodshed once again. And not just
in Port-au-Prince. In Gonaives, Cannibal Army leader
Amiot "Cubain" Metayer had just turned up dead with his
eyes and his heart carved out. His followers were sure the
National Palace was behind it, and they were now quickly
taking control of villages and towns in an organized sweep
toward the capital. A second group calling itself the Liberation Army was on the move as well. It was mainly made up of
former army officers who'd been biding their time ever since
Aristide dissolved the military in 1994. If these hundreds
of fighters could reach Port-au-Prince and link up with the
hundreds of thousands of protestors, they might just topple
the Aristide government.

Didi was right. It was time to get back to Haiti.

"I can't go right now," I said.

"I'm sorry?"

Her surprise didn't surprise me. It was the first time with NPR that I'd uttered such words. And at the height of a crisis on my watch. This was highly out of character, but the circumstances were highly out of the ordinary.

"I can't go," I said. "I need at least a couple of days. Anne nearly lost our baby today."

"She's pregnant?"

"Yes, six weeks. She developed some bleeding last night and has to stay in bed."

"Well."

"So that's my situation."

"Haiti is on the verge of collapse."

"Yes, I know."

"Loren is going to want someone on the ground."

"If we can hold off a few days, then I can go."

"It looks unlikely that the situation is going to calm down."

"It certainly does look grim."

"You know, my husband was a cub reporter being chased by lions in Africa when our first child was born."

"I did not know that."

"That's the life we'd chosen."

"I'm trying to think who might be able to cover for me for a few days."

"I don't know who that could be."

"We can narrow the list," I said, "by crossing out anyone who's worried about losing their baby today."

Didi didn't say anything. It was clear to me that she was

more concerned about Jenkins's potential reaction to my refusal to go than about why I was refusing. In some ways I couldn't blame her; Loren was a demanding, sometimes prickly boss who expected results 100 percent of the time. It made working for him difficult at times. But it was also what made NPR's foreign desk so productive and well respected.

"Look," I said. "I'll call Loren myself if you'd like."

"If you want," she said, accepting the out.

I hung up and went downstairs to the couch.

"They want me back in Haiti as soon as possible. That is, on a plane to Miami tonight."

Anne started to cry.

"But what am I supposed to do here alone? I'm supposed to stay flat all the time. What if we lose the baby? Didn't you tell them?"

"Shhh, hey. Yes. And I've already told them I can't go."

"What did they say?"

"I still have to call Loren."

I went back upstairs.

"Yes, Gerry."

"Loren, look. I just spoke to Didi, but I can't go to Haiti right now."

"What do you mean?"

"My girlfriend nearly lost our baby today. She's laid up and needs me to take care of her for a few days."

"Well, *fuck!*" he screamed.

"Listen . . ."

"Now what am I supposed to do?!" he said. "Now I've got

to dig up another fucking reporter from somewhere else! Where the *hell* am I going to find someone else?! *Fuck!*"

"I'm sorry, Loren. You know that under any other . . ."

But I didn't bother to finish my sentence. The line was dead. Jenkins had hung up on me.

Downstairs Anne was talking on her cell phone. I felt like my head was going to explode. Jenkins wasn't forcing me to go, but he'd made his displeasure more than clear. On top of being worried sick about Anne and the baby I now felt concerned for my job. Jenkins had fired people for less. When Anne saw me, she hung up. She stopped crying and looked more composed.

"Go."

"What?"

"Go to Haiti."

"But I've just told them I'm not going. I'm not going to leave you like this."

"I spoke to Olivier. He can come over and help me." Olivier, a young French teacher from Réunion Island, was Anne's best friend in Mexico. "He'll stay here until you come back."

"Are you sure? Jesus. What if something happens?"

"Something could happen whether you're here or not. If the radio really needs you, you have to go. I won't be alone."

I climbed back up to my office all the more disgusted for Anne's grace. First, I called Jenkins.

"I'll go," I said.

"Glad to hear it."

I called Didi back.

"I'm going," I said.

"So now you're going."

"I'll be there by midday tomorrow. And if Anne and the baby are okay, then in a couple of weeks we'll be going to Rio."

"I suppose that will depend on the situation in Haiti."

"Rio," I said. "Then Ilha Grande, a couple of hours to the south."

I hung up. The Brazilian vacation was long scheduled and now, more than ever, I planned to go. If my editors couldn't cut me slack, I was going to take it.

It was January 28, 2004, and we were witnessing another round of protests in Port-au-Prince. I'd been on
the ground for three days now and Anne was still okay. The
bleeding had stopped the evening after I left and hadn't resumed. In less than a week, hopefully, she'd be back on her
feet. Here on the ground everything was different. Over the
last four years I'd developed a certain intimacy with Haiti's
tensions and its struggles, but never this. What we were
watching now was intimacy's undoing.

A small group of medical students was trying to jump-
start a larger demonstration as they ran laps through the
streets around their downtown campus. Some of them
held up banners calling for President Aristide's resignation.
These were the same students who, two months earlier, had
witnessed pro-Aristide thugs shatter the legs of the university's rector with clubs during a protest against his dismissal
by Aristide. Now their movement, backed by a business elite

bent on toppling one government in order to install another more amenable to free-market liberalism, seemed unstoppable.

As the protestors ran circles around the block the number of spectators grew. Some cheered them; others began to heckle. The students continued, sweating and singing. One stretch of their run brought them right past a crowded Aristide stronghold downtown, and from there it seemed that some kind of counterattack might erupt. Soon Haiti's riot police, the armor-clad CIMO, arrived on the scene. As the morning wore on, the crowds watching from the sidewalks grew bigger. Every three or four minutes the dozens of medical students would round the corner again, working themselves into an ever more agitated state, which had a similar effect on the onlookers. At some point the CIMO decided they'd had enough. They physically tried to stop the protestors, who eluded them and ducked into a building on the medical university campus, locking the gate behind them. It seemed that the situation was now contained, but the CIMO began firing teargas to flush the students out.

I was standing some fifty meters from the contingent of CIMO as they launched teargas canisters over the gate. I hadn't been paying much attention to the nearby crowd until I suddenly felt a sharp pain in the back of my left leg. I turned around just in time to dodge a second chunk of cement flying toward me. The large crowd on the nearby street corner—young men in tattered T-shirts, heavy-set women selling small balloonlike bags of drinking water—regarded me with stony gazes or not at all. Whoever had thrown the stones wasn't showing himself. But I didn't want to wait around for more trouble. My fundamental problem was that I had violated yet another of the cardinal rules of my hostile environment training program: I'd let myself get separated from the rest of my colleagues and had no sense of the allegiance of the people around me. I had no idea whose camp I'd wandered into.

The CIMO were now busy not only firing gas canisters but dodging the ones that the medical students were throwing back at them. The curtain of smoke before me was nearly opaque, and I didn't want to risk trying to cross through that crossfire. But by staying put I was risking a rock fight, pitting me against half a neighborhood. You could feel it coming. For a moment the gas dispersed enough, and I could see a knot of foreign reporters through the haze some hundred yards away. I would certainly feel safer on the far side of things—if only I could get over there. Running around the block was out of the question. I tried to call my driver, who was parked somewhere nearby in his truck, engine running. His instructions were to come scoop me up on my word. But the phones weren't working. Finally I decided I had no choice but to make a dash for it across the smoky line.

I took a breath and held it and ran squinting against the sting of the teargas. The police didn't notice me, and I made it without incident into the sun beyond them. But the safe side of things wasn't any better.

I came immediately upon a crowd of frantic Haitian

parents screaming and banging on the large green metal doors of an elementary school, just across the street from the university. The teargas was wafting into the school, and mothers and fathers were trying to get their kids out and away from the scene. At first the terrified teachers wouldn't open the doors despite the frantic entreaties outside and the wailing of their charges around them. When they finally relented, dozens of coughing parents stormed in and then out, carrying their kids or leading them by the hand as fast as they could down the street. One uniformed schoolgirl in a white blouse and plaid skirt reached the street and fainted from fear. She fell like a kite to the ground. A young woman nearby half-dragged her across the street and back into the school. Teargas continued to fly. The loud reports from the CIMOs' teargas launchers made me flinch. Another group of young men went sprinting past us. There was more firing and suddenly the crowd outside the school was yelling again. This time they were calling for an ambulance.

In the center of the confusion and smoke sat a young man in jeans and a dark T-shirt. As the crowd turned and dashed and swayed, he sat still, his legs stretched straight out in front of him and his torso inclined slightly forward. He was looking calmly to the left and right as colleagues waved their arms frantically. I saw that he was sitting in a pool of blood, but I couldn't see any wounds on him. As I neared, another young man lifted up the back of the fellow's shirt exposing a weeping gash that made me think of the damage a harpoon might cause to a whale. The wounded man, a reporter told me, was one of the protestors. He had been struck directly in the back by a teargas canister as he'd run past the same CIMO line I'd just crossed. The canister had ricocheted off the man's back, but not before opening up a flap about five inches long just behind his right shoulder blade.

Purely by chance, an ambulance manned by Chilean medical volunteers was idling nearby. It quickly arrived and whisked the wounded student away, taking him to Port-au-Prince's only hospital still functioning, the private Canapé-Vert Clinic, up the mountain some two miles from the protest scene.

The downtown march abruptly ended, but as word spread that a protestor had been wounded by the police, a huge crowd took off jogging for the Canapé-Vert Clinic. I finally reached my driver on his mobile phone and he sped over. I dove in and we raced up the mountain road ahead of the surging crowd.

The Canapé-Vert Clinic was a private hospital, perched on a small, well-tended hillside overlooking a major intersection on the road from Port-au-Prince to the upper-middle-class enclave of Pétionville. In the center of that intersection was a small grassy park with a stepped amphitheater. When we reached the clinic there were hundreds of people already there.

We went inside and learned to our surprise that the wounded man had died; his injury was worse than anyone had suspected. The teargas canister had not bounced clear on impact. It had perforated the man's body and lodged inside his chest cavity.

"We rushed the patient into surgery," a surgeon told me, "but once he was on the table the gas canister began to sputter and release its contents. We had to flee the operating room because of the gas."

"And the patient?"

"You can imagine," he said.

He took me into the operating room so that I could see for myself. It was indeed the same man, shirtless now and prone on a stretcher, his chest bound with white gauze. The doctor rolled him over briefly and showed me the entrance wound and the incision through which they'd finally removed the canister once it had discharged its load. The canister was there in a metal dish. Blood and burned tissue caked into its metal casing. We went back out into the crowded hospital corridors.

Out on the plaza the anti-Aristide crowd was gearing up for a fight. Several men were piling stones in the street like cannonballs, while others hefted them in their hands. At the top of the park a large group of Aristide supporters began congregating. Soon insults were flying across the plaza, unintelligible shouts echoing across a no-man's-land of about two hundred yards.

I had wandered out into the amphitheater with a photographer for Polaris Images named Timothy Fadek. The sun had just set behind the hillside and we were now between the opposing groups—another no-brainer no-no from Commando School 101. In a matter of seconds, if the two sides surged, we might easily become a reporter sandwich. The only explanation I have for such recklessness is that sometimes you just don't see yourself among the images unfolding. You think: I am observing all of this from someplace safe. I am a recorder, not a participant.

Such foolishness. Some anti-Aristide protestors were creeping farther up the street and igniting tires in the road nearby. It seemed they were trying to dissuade their enemies from storming the clinic. We should have taken more precautions as the light grew dim.

Suddenly a truck burst through the smoky roadblock. It was a CIMO transport vehicle packed with police. Several more CIMO were running behind the truck, their rifles raised. They had not been provoked, yet they were speeding straight for the clinic. Tim and I quickly realized that we had better get down the hill fast; once the police passed us, the angry Aristide supporters would take the plaza. Of late they'd decided that the foreign press were part of the conspiracy against their leader—Aristide himself had hinted at such during a recent radio broadcast—and their increased hostility toward us had grown palpable. We took off running.

Luckily the police vehicle slowed a bit and we reached the road just ahead of it. The huge crowd was scrambling up the clinic driveway and through the surrounding bushes and trees, frantic to get away. We were making a beeline for the driveway several yards behind the tail end of the crowd when gunfire suddenly erupted behind us. The CIMO were firing live rounds indiscriminately into the air. I glanced over my shoulder and saw multiple muzzle flashes. The police truck accelerated, and it became clear it would overtake us before we reached what I believed was the safety of the clinic. We dove instead for the far side of the road, throwing ourselves prone in front of the locked gate of a private house just short of the clinic driveway. The CIMO truck thundered past us and turned into the clinic entrance. The CIMO on foot were firing like mad now, just feet from us. It sounded like a fireworks show. Tim was trying to shoot images, which greatly disturbed me. I didn't want the CIMO to notice us for fear they might mistake us for protestors, even if just for a split-second, and direct their bullets at us. The CIMO truck roared up the driveway but only made it halfway before getting stuck under a cement and iron archway. The CIMO jumped down and began running across the clinic grounds. From the street we could hear screaming and more shooting. The CIMO truck finally dislodged itself and backed out to the road again, then went tearing farther down the hill. The Aristide partisans were now crossing the plaza toward us.

We ran up the clinic driveway. By the time we reached the hospital there was almost no one outside. Everyone had taken off running. But some unlucky few had tried to take refuge inside the clinic. Unlucky because the police had decided the medical facility was not a safe harbor. As we passed one clinic window I saw a CIMO running down a corridor, weapon raised.

The gunfire had abated some. I ran to the main entrance

and saw that someone had swung shut and locked a barred metal security door. Just on the other side of it, on the white tiled floor, two young Haitians lay crumpled, one on top of the other. I saw blood beneath their bodies. The youth on top was lying faceup, his mouth open slightly, his eyes shut. I could not reach them, but I watched them for some ten seconds, hoping to see some movement, a subtle inhaling, something. But they appeared dead.

In that moment I heard somebody shouting my name. It was my fixer. He was parked in a dark corner of the clinic grounds, engine running, truck pointed toward a back exit. "Let's go!" he screamed, "Now!"

I ran and dove into the truck. The Aristide crowd was now climbing the steep driveway. Fadek, right behind me, jumped into the pickup's bed and we went hurtling down the mountain. I couldn't believe what I'd just seen. Why had the police provoked this clash? The government had been enduring weeks of such protests, and without doubt they were having a destabilizing effect on the country, but storming the capital's only functioning medical clinic was surely going to backfire on Aristide. Had he ordered the attack or had the police made their own call?

We raced back down the hill to the Oloffson Hotel, just blocks from the National Palace, and I quickly called my editor and filed my story. Then I had a drink at the bar and collapsed into bed.

I was awoken before dawn by the ring of my mobile phone. "Hi, Gerry," began a familiar voice, "I just wanted to thank you for publishing the photo of the charred man on the NPR website."

"Who is this?"

"Michelle, from the palace." It was Michelle Karshan, Aristide's international press spokeswoman. She was referring now to a man who had been doused with kerosene and set alight, suffering third-degree burns over 70 percent of his body. I had stumbled upon him in the entrance to Port-au-Prince's main hospital while seeking interviews two days earlier with what staff dared show up there for work.

The man was barely alive, naked and writhing slowly on a gurney in the emergency room. His skin curled off his limbs like loose birchbark. Only one doctor was working in the entire hospital because of the unrest.

"In recent days thugs have been roaming the halls," he told me, "stealing whatever they can and even raping patients. None of my colleagues will come in." As he was talking his eyes suddenly drifted over my shoulder and went wide. I turned around. The burn victim had somehow found the strength to sit up. We approached him and he began whispering to my fixer.

"A gang of men," he said, "a gang of men set upon me while I was working in my yard. They burned me because I support Aristide."

"Is he sure?" I asked.

"Are you sure?"

"They told me that I must pay for supporting Aristide," he said, as the doctor eased him back on the gurney.

I took pictures of that near-dead man, just stuck there without any medication to ease his suffering, and sent the images to Didi in Washington. They were published on the website the following day. Hence the call from Karshan.

"It's inappropriate for you to thank me," I told her, annoyed. "I don't suppose you're calling to explain why the police shot up protestors last night at Canapé-Vert."

"What!?"

"During the protests at the hospital. I saw what looked like two very dead people inside the clinic just after the CIMO entered the clinic firing their guns," I said.

"Fuck!" she shrieked and promptly hung up on me. It seemed I was having this effect on people lately.

That morning my driver and I returned to Canapé-Vert to find out exactly what had happened the evening before. When we arrived life seemed to be back to normal. Patients, guests, and a few doctors and nurses brave enough to come to work milled about inside. I began asking staff if they had seen or attended to two Haitians lying prone in the entrance the night before. Nobody seemed to know what I was talking about. And judging from the rather cold reception I received, it appeared that nobody really wanted to talk either.

I left the clinic none the wiser. Outside it was already very hot. The sun-drenched street in front of the clinic was choked with cars and colorfully painted pickup-truck taxis going up and down the mountain. As we neared our truck a middle-aged man in a green jumpsuit came running up to us.

"Wait! Wait!" he hissed. "The police took the bodies away last night."

"Who are you?"

"A janitor at the clinic. I was too afraid to talk inside."

"Where did they take the bodies?" I asked.

"Who knows," said the janitor. "They made us mop up the mess, to get rid of it."

"Who?"

"The CIMO," he said.

"And then what?"

"That's it," he said, then hurried away. He would not speak on tape. I was thinking about taking a trip to Port-au-Prince's morgue to see if I might recognize the two bodies when my phone rang. It was Michelle Karshan again.

"Hi, Gerry," she said cheerfully, "I just wanted to let you know that the Palace sent a security detail up to the Canapé-Vert hospital this morning to investigate your claims about the bodies. It turns out there weren't any."

There weren't any?

"I saw them with my own eyes, Michelle. They were lying just inches away from me. So did a photographer. A TV cameraman who I didn't know was there as well."

"I have no reason to doubt the Palace," she said. "There must have been some confusion. I understand it was all very hectic."

I was at a loss. All I could think to say was "Okay, Michelle," and then I hung up. When I told my fixer what Karshan had said, he was not surprised.

"If the Haitian national police kill people in a clinic, then force the staff to clean up the evidence, the government obviously doesn't want people talking about it. So why would someone talk about it when Palace thugs come by the next day?"

He was right. We drove down the mountain to the morgue, on the off chance that the police hadn't just dumped the bodies outside somewhere. But we did not find them there among the rising number of dead.

"Hey! Hey! You're going the wrong way!" The woman in the white dress with the suitcase balanced on her head was shooing along four small children in my direction. They had just climbed around a barricade of trees and stones that blocked the main road into the town of Saint-Marc, an hour north of Port-au-Prince. Behind her, black smoke rose into sky. "You're going the wrong way," she yelled again.

I'd driven out to Saint-Marc after the Canapé-Vert incident because Saint-Marc was now the front in the rebel advance on Port-au-Prince. When the woman reached me she grabbed me by the arm. "*S'il vous plaît. S'il vous plaît.* Turn back. That's where the fighting is!" She pointed back across the barricade.

"What's going on in town?" I asked.

"We have to send the kids away," she said, her voice filled with fright. "There's too much shooting. We can't live in this environment. It's a complete state of lawlessness. People are on a rampage, grabbing people off the street and shooting them."

More people came around the barricade, carrying what belongings they could.

"The town is about to fall to the rebels!" a man on a bicycle shouted, without stopping.

I climbed around the barricade and flagged down a teenager who'd been ferrying people out of town on his moped. He took me toward the center of town where a group of police had set up position. In the very near distance smoke was rising from multiple locations. A police helicopter swooped past, bringing in reinforcements. The police told me to turn back, but just up the road an even larger group of men in civilian clothing was milling about. We sped over to them and jumped to the ground. They were armed with pistols and rifles of every make and use, and they surrounded us nervously.

My phone was ringing. It was Anne.

I turned it off and introduced myself via my fixer. I didn't ask to interview them; I just turned on my deck and pointed my microphone at them. It was not a time for politeness or for showing hesitation whatsoever.

"Who are you?" I asked.

"We are members of Clean Sweep," said a man who was wearing bullet belts crossed over his chest.

"What is that?"

"We're the last of the government supporters in Saint-Marc," said another man, his eyes glued farther down the road, toward the trouble that was coming. "The last. The houses of the Lavalas partisans here have been burned down. All of this is happening and the president is still in power. What will happen if he's forced to leave? What will we do then? What will happen to us?"

Another man in a Michael Jordan tank top interrupted, smirking.

"We will defend our president with sticks and stones against the rebel guns," he proclaimed. "They are the ones who are armed, not us." As he spoke he was hefting a grenade in one hand like a dark gray snowball. But his bravado did not reflect the collective mood amid that smoke and erratic gunfire; all around us villagers were now scurrying past, carrying mattresses, musical instruments, suitcases, farm tools. Everyone was fleeing for the capital. But they'd find little protection there. In peaceful times Port-au-Prince could not have handled a surge of refugees. Now the situation was dire because the capital's most important north-south supply line for food and fuel had just been cut off. I could confirm that. I was standing on it. The road through Saint-Marc, Haiti's National Highway 1, was barricaded at one end of town and had fallen into enemy hands at the other.

If the Clean Sweep fighters thought their situation was desperate they only needed to consider the plight of the gangs holed up in Port-au-Prince's largest slum, Cité Soleil. These young men had thrown in their lot with Aristide too, and now that his future was uncertain theirs was even grimmer.

For as much as Cité Soleil's endless, unmarked alleys offered protection for gang members, the slum as a whole had already become their inescapable tomb. With Aristide on the defensive, the gangs' once-cowed enemies were beginning to prowl Cité Soleil's perimeter. And now, one leader told me, they had reason to fear Aristide himself.

I'd known the kid for years. His name was James Petit Frère; some people called him "Bily Iron Pants," or just Bily, which was how I knew him. He'd grown up partly in an orphanage run by a foundation that Aristide had set up for street kids. As such, he knew Aristide personally. Early on this had added great privilege to his life, but now it was bringing him peril of the highest order.

As a young teenager Bily had befriended a French photographer who used to come around Aristide's foundation. From time to time she would bring Bily to the Oloffson Hotel for lunch or an ice cream: that's where I first met him. You could see how bright he was and how much he admired all that sitting on the Oloffson porch implied. Just across town from Cité Soleil but a world away.

But the truth was that Bily, now grown, had a foot set firmly in both worlds. When he'd left Aristide's orphanage to return to Cité Soleil the president did not forget him. Through the national police, Aristide had kept money in Bily's pockets and, when necessary, weapons in his hands. Capitalizing on the slum youth's desperate devotion and loyalty, Aristide had turned Bily into one of the leaders of the *chimères*—the unofficial henchmen upon whom he could call to protect him from threats. In return, Bily told me, he and his friends had been promised that they, and Cité Soleil itself, would have brighter futures. Jobs, roads, sewers, light.

The last time I met with Bily, just after returning from the Saint-Marc front, he was decidedly less optimistic about his future. And he'd grown suspicious of his benefactor's hand in it. For the interview I wanted to do he asked that I give him a false name, out of fear for his safety. So in my NPR story I called him simply François. I use his given name now because he is dead.

When I went to see him that last time, it was early February and he and his men were preparing for war—literally against everyone and anyone—inside their shantytown stronghold of half a million people. They no longer knew whom to trust. They'd been fighting for years against Aristide's enemies and slum upstarts trying to muscle their way in, but now they had an even more fearsome foe.

"The police are coming for me," Bily told me. "There's no way Aristide can let me and my boys live. We know too much. We done too much stuff behind the scenes."

To even enter Cité Soleil now you were best advised to fix an appointment. Showing up unannounced could get you shot at or shot up. Bily's boys were guarding the entrance and had orders to stop the unannounced and the suspicious. So my fixer and I arrived at the slum's entrance by car, then parked and waited for one of Bily's henchmen to retrieve us. We walked in, under escort, through the running sewage and past the children playing on mountains of garbage and

the pigs foraging in the hot mud for feces and scraps. This
was the Haiti you often saw in newspaper photos, the Haiti
on its knees, at its least humane. It was also the Haiti that
most Haitians hated to see publicized. The nation was so
often characterized, in words and images, as the "poorest
in the Western Hemisphere." This portrayal grated on the
people and contributed to the country's collective desolation.

And yet here it was, Cité Soleil, dismal and unchanged year
after year, generation after generation. Given the uncertain
security situation, we were forced to wait for Bily for a couple
of hours in what passed for a small plaza, watching a group of
men playing dominos in the shade of their cinderblock huts. It
was hot and it smelled. At one point a drunken man appeared
and tried to pick a fight, first with one player, then another.
Finally he upset the crate upon which their dominos rested.
Suddenly fifteen men were on their feet screaming, and that's
when Bily appeared at last. Just at the right moment, like a
cartoon hero or its villain—that was always the question.

His pistol was already drawn and he put it to the head
of the troublemaker as he removed a gun from the man's
waist. Then he said a bunch of things to him in Kreyol that
convinced the drunkard that it was wiser to go away. Only
when the dominos game had resumed did Bily turn to greet
us. He wore a big, worried smile.

"Follow me," he said.

Nineteen years old now, tall and lanky, in drooping pants
that hung even lower from the weight of his pistol, Bily was
the cops and the courts in his section of Cité Soleil. The po-

lice might be hunting him now, but in the days of deeper understanding they rarely came in here. And when they did, it was usually to meet with Bily or one of the other five or six neighborhood gang leaders to work out arrangements with them. The police, Bily had told me on more than one occasion, served as the conduits between the gangs and President Aristide himself.

Bily led us on a winding walk that took us past, among other things, a half-built community center and a concrete, fenced-in park where kids were playing soccer. People scattered like pigeons as he passed, out of fear or respect or the two intertwined.

"I'm building all of this with my money," he told me, pointing to the community center. "You see? I am trying to change Cité Soleil. If I am killed, one day the people of my city will remember me. They will talk still about Bily."

Now I understood. Bily's pre-interview walkabout was about establishing his legacy. He must have considered himself as good as dead already—as if all that remained to write was his obituary.

"Where do you get the money for all of this?" I asked.

"From the chief of police," he said. "The chief pays me to back Aristide up. When he needs me, I get the people out."

The tour of Bily's good works ended at the port, and we went into the main offices there. When the manager saw us and who was leading us, he gave up his desk. Bily sat down behind it and put his gun down and rubbed his face. He had grown much gaunter and tougher in this *chimère*'s life.

For two years now he'd been running a large section of Cité Soleil, having taken over from his older brother, who was in jail. Bily told me that there had been doubts at first as to whether he could fill his brother's shoes until the day he executed a rapist in front of the victim, the victim's mother and father, and half the slum.

"I didn't like that, sir," he told me. "But the girl was bleeding a lot. I told the guy, 'You're not gonna do this anymore.' Then, BOOM. Then I gave some money to the girl to go to the hospital. And everybody was happy. Everybody was clapping for me and said, 'This is good.' And since then no more rapes have occurred. At least not in my area of control."

And since then, he told me, no one had stepped forward to challenge his authority. "But now," he said, "the same people who give me the means to power are coming to take it away." Aristide's national police, the same ones who for years ferried Bily cash and bullets and orders from on high, were pursuing him. If Aristide was in danger of losing his grip, his secret security force must be eliminated. They were the proof of the lengths to which Aristide had gone to fend off his own numerous enemies: members of the disbanded army, the heavily armed private guards of the nation's few oligarchs, rival political factions in Port-au-Prince and elsewhere, and displeased elements of the "international community" stubbornly bent on his demise.

Earlier that week during a press conference at the National Palace I had asked Aristide about the gangs' claims. The president denied everything the gang members had been telling me and other foreign journalists—including that they were now palace targets.

"This is a way for those who want to criticize, to criticize," the president said. "But as I said, it is false. We have the responsibility as a government to disarm in a legal way all those who have illegal weapons in their hands."

When I asked Bily if he'd hand his weapons back in to the government, he laughed.

"Aristide would have to live up to his side of the bargain first," he said, "by bringing something new to this slum."

I do not know the Kreyol word for dignity but Bily said it. Dignity, he said, for all of these jobless, hungry families forced to shit among pigs on the steaming shoreline and the young falling sick from it.

"If Aristide doesn't make a decision to change something with us, we can be mad with him too. And then he better watch out. Because we want a new life in Cité Soleil. To have an education. To change the situation with these guns. We don't want to use guns no more. We want to use the book."

We went back outside after a while where Bily's boys were standing guard. He and I walked alone to the edge of a pier and leaned against the rail overlooking the sea. He pulled his gray, snubby revolver from his back pocket and looked at it closely, then at the water below. Suddenly I knew that he was going to throw it into the sea. Given what he'd just been telling me. But I didn't know anything. He just opened the chamber and checked his bullets then snapped it shut and put his hand on my shoulder.

"The Oloffson Hotel," he said, shaking his head. "You know, I've been sleeping in a different place each night for a year now. Every night, blam! blam! blam! Against the cops, against everyone.

"But now I don't care if I die," he said softly. "I am ready."

Here was this brave, complicated vassal. Fallen from favor, ever dispensable and encircled now in his miserable fiefdom. And all he could do was wait for the end and make his peace with it. He could put a brave face on it.

"The day someone comes for your power," he said, aiming his gun at the benign sky, "you *better* be ready."

That was the last time I saw Bily Iron Pants. When Aristide fell, so did the secret network supplying Bily and his boys with their sustenance of munitions. They were left to defend themselves against an onslaught of vengeance seekers with no shortage of bullets. Within months Bily was shot, then thrown in jail. He managed to escape, but a short time later the police recognized him again and killed him on the spot.

A few days later, back in Port-au-Prince, I was covering the largest anti-Aristide protest yet. The police and marchers were out playing hot potato again with teargas canisters. Several days had passed since my visit to Saint-Marc. I was on my cell phone to Anne.

"No! It's a peaceful march!" I was saying. Then I saw that Didi was trying to get through. It was hard to talk amid the chaos.

"If you've still got someone to cover for me," I yelled over the din, "then I'm out of here the day after tomorrow!"

"Well, Kaste *says* he can cover."

"Okay! Kaste it is, then! I'll get in touch with him!"

Two days later I met Martin at the airport in Port-au-Prince. I'd never seen such crowds there. Hundreds if not thousands of Haitians were trying to get on planes. The parking lot was a madhouse of screams and luggage. Martin was one of just a handful moving in the opposite direction. I sat on the roof of my fixer's car and spotted him easily as he exited the baggage claim hall onto the street. He had never been to Haiti, and I could understand why his eyes were like that.

"You'll be fine," I told him, introducing him to our fixer. "He'll take you to the hotel and get you situated."

"No problem," he said, taking in his surroundings.

"I appreciate your coming," I said.

"Enjoy Brazil," he said.

I planned to. Although it would be a bit strange now. Before Haiti had gone critical, the plan had been for Anne and me to stay for a night or two with Martin and his wife, Amy, in Rio, then head for Ilha Grande. Now, with Martin covering for me, there would be an awkward moment visiting with Amy alone, as images of Haiti's unrest beamed across her TV. From the outside it did look difficult. But I had covered for Martin on a few occasions, I told myself. Martin and I talked strategy for a while there in the airport parking lot, then I wished him luck and fought my way into the terminal for my flight to Mexico City.

———

Anne and I were on what might be described as the perfect beach—a white crescent one mile long, hemmed by the sea and a thick green head of jungle—when the owner of our rented bungalow came by to give me a hand-scribbled note with a phone number. "Call Louro Chenkis."

"Loren, Gerry."

"I need you back in Haiti now."

"Not a problem," I said. "I brought all my gear with me just in case."

I had not seen a newspaper in a week.

"It looks like Port-au-Prince is going to fall," Jenkins said. "That means the country."

"I'm on it," I said.

"Where are you exactly?"

"Ilha Grande."

There was a long pause.

"I would have thought you'd want to be there. The country is in flames. This is the biggest story on your beat since I hired you. I'd have thought you'd want to cover it."

"What's paramount to me," I said, "is that NPR has the story covered. Whether it's me or another competent reporter is secondary. I told Didi before I left that if you needed me to stay, I'd stay. Or if you needed me back there, I'd cut my vacation short."

"Well, I don't know how long it's going to take you now."

"I'm on my way."

Anne and I were on the next ferry to the mainland. I left her with a decidedly worried-looking Amy Kaste in Rio and sped to the airport to get on the standby list. That night I made it as far as Miami, where I slept on my bags in the terminal. In the morning I was first in line at the American Airlines window. Behind me was Scott Wilson, from the *Washington Post*, and another reporter from the *Boston Globe* who'd never been to Haiti. All of us were hoping to catch the early flight to Port-au-Prince. But it wasn't going to happen. American Airlines had canceled service to Haiti for lesser reasons than civil war.

"Today," an airlines representative told us, "there is no way we're going in."

"Tomorrow?" we asked.

"You don't seem to get it," he said.

With direct flights into Port-au-Prince out of the question the three of us did the next best thing. We fought our way onto a flight to Santo Domingo, the capital of the neighboring Dominican Republic. From there we might get into Haiti by air or over land. We had no plan except to try to get closer and closer. But our hopes were crushed in the Santo Domingo airport. There we learned that the Dominicans were exercising the same caution as the Americans. All flights into Haiti had been suspended.

"It's a civil war over there," an airport worker told us scornfully. "Once again."

And the road?

"You can drive to the border," he said, "But they won't let you cross."

By late that night there were nine of us brainstorming an entry plan. Eight journalists and an American in a business suit named Reggie Barnes who'd been wandering the same halls all day, two cell phones to his ears. Eventually he'd overheard us and introduced himself. He said he was the "attaché" of a great man by the name of K. A. Paul. Reggie was middle-aged and southern and had thick white hair. He could have passed for a congressional aide or even a congressman himself.

"Do you not know the man I work for?" he asked when no one reacted to the name. "K. A. Paul is one of the most important humanists alive," he said. "He's the world's most popular Christian evangelical. And he is on his way to save Haiti from terrible destruction."

"Does he have a plane?" I asked.

"Yes, he does," Reggie said. "Dr. Paul owns his own 747. *Global Peace One*'s the name. At this very moment it is being loaded with provisions for the suffering Haitian people."

"Here?"

"It's in Miami."

"Could it pick us up?"

"That is what I'm working on, friend. There'd be room for all of you." Reggie lifted his two cell phones again but could not seem to reach a soul.

"Damn it," he said to no one in particular. "Where is that plane?"

"I'm sorry," said Wilson. "Did you say K. A. Paul was a humanist?"

"He is the most popular Christian evangelical in the world. Dr. Paul was born in India, where he has saved countless orphans and widows from misery and death through his Global Peace Initiative. He personally brought President Charles Taylor of Liberia to God and convinced him to relinquish power in the name of saving innocent lives. In Haiti—if we're not too late—he is going to pray with Guy Philippe that further bloodshed may be avoided."

Guy Philippe was a former Haitian police officer and the son of a wealthy family from the country's south. He was now leading the Liberation Army in its advance on the capital. He had become, in effect, the guy we were now racing against to the National Palace. How had this K. A. Paul guy gotten in touch with Philippe? We were all looking at each other for the insight that we all clearly lacked.

"Do you know Gandhi?" Reggie went on. "Martin Luther King? Gentlemen, you have stumbled upon history in the making. With any luck I'll be introducing you to Dr. Paul shortly."

"I'll interview him if he can get me into Haiti," I said. "And I'll interview him a second time if he stops the rebels."

"When you pray with Dr. Paul you understand why world leaders roll out the red carpet for him," Reggie said knowingly. Then he recalled his mundane task at hand and started working his phones again.

"Goddamn it, where is that goddamned plane!"

"I think we need a plan B," Wilson whispered to me. We ran over to a row of pay phones and began calling every private charter company in the Yellow Pages. Wilson was on the line with one when he cupped the receiver and yelled out to us.

"Listen up, everyone! I've got a pilot on the phone who says he'll fly us in for six thousand dollars! The more we are, the less it'll cost each of us." Everyone looked at Reggie, pacing back and forth, dialing, hanging up, cussing to himself.

"Or we wait for the humanist," Wilson said.

The twin propeller plane sat ten to twelve people. I made sure I sat as far away from Reggie as possible. It was barely light outside when we took off, and below us I could just make out Santo Domingo as we left it behind and whined toward the Haitian border. Within twenty minutes we were high over the Sierra de Baoruco mountains as they rose, green and irregular, under the first rays of sunlight. Then we reached the Haitian border.

You knew you'd arrived because you could see clearly where the Dominican forests stopped and the Haitian desert began. "The mountains are showing their bones," a Haitian charcoal vendor once commented to me. Below us, along the peaks and valleys of Haiti's once-lush Massif de la Selle, virtually no trees were standing. From on high, the landscape looked like the skin of a plucked chicken, pale and ruined. After another half hour our pilot touched down at the darkened Port-au-Prince airport.

"I don't turn the motors off!" he reminded us. When the plane had taxied to a stop, he popped the door and lowered the ladder and we climbed down onto the runway, dragging our bags with us. Then the door shut and the plane hung a tight turn. We watched it lift off as Reggie was trying to get his phones to work.

"Whew," sighed the reporter from Boston. "At least the hardest part's behind us now."

Wilson and I exchanged glances. We approached the terminal.

Inside it was dark and empty, and felt suspended in time like a sunken ship. We walked through the unmanned customs gate, past the ticket windows, and down the short hall toward the baggage area. A welcome sign fashioned from metal that read WELCOME TO HAITI, CELEBRATING 200 YEARS OF INDEPENDENCE hung dented and half-fallen from a wall. On the street a couple of cars were idling, including my fixer's.

"You're late," he said to me, annoyed.

"You're kidding," I said.

We managed to fit everyone into the two vehicles, then gunned it for the road. Fires were sputtering here and there, mostly car tires lit hours earlier and some buildings as well. But it had rained heavily during the night, dousing most

of the blazes. It was just barely early enough still for the
streets to be devoid of life. We took a back route at very high
speed to the Montana Hotel, passing absolutely no one on
the roads. When we reached the hotel, though, there was
lots of activity. Private security forces were guarding the entrance. Reporters and well-dressed foreign NGOs were already milling about the parking lot. We sped past them and
checked in quickly. I knocked on Martin's door. He seemed
glad to see me.

"Welcome back."

"How'd it go?"

"We've been busy," he said, yawning. "Though it was hard
to leave the hotel grounds yesterday. Somebody dumped a
bunch of bodies just beyond the entrance and people were
saying anyone who left might get shot."

"That's grounds for staying put," I said.

"But we went out."

"And?"

"It was hairy. People were shooting each other and looting stores and banks and food warehouses. Everybody's really freaking out."

I called Didi.

"The two of you work out your strategy for today and get
back to me," she said.

I told Martin to take a rest day at the hotel, but he refused. I called Anne in Rio to let her know I'd arrived and
that anything she might see on TV was being blown way
out of proportion. Martin and I had a quick breakfast and
ventured out downtown.

It was the quietest morning. The rains had apparently cooled collective tempers somewhat. But the situation for Aristide looked even more precarious. Port-au-Prince, like its Cité Soleil slum, had become cut off from the world around it. By air, by sea, by land. Food supplies were dwindling; prices were rising. The food crisis was in part what had sparked yesterday's citywide looting. But there were other pressures. A person at the U.S. embassy who would not speak on record confirmed rumors that the Cannibal Army, now joined up with the Liberation Army, was already on the outskirts of the capital.

Publicly, the United States was now directly pressuring Aristide, questioning the wisdom of his leadership and his capacity to lead. France's foreign minister had grown even more blunt, calling on Aristide to abandon the National Palace. So far such diplomatic broadsides had only served to rile up Aristide's most militant supporters, many of whom were heavily armed and had been patrolling the capital in pickup trucks in recent days, flanking the ever more desperate pro-government marches. I was sure Bily Iron Pants was among them, throwing his support behind the president in one last effort to save Aristide—and himself.

That afternoon there was another pro-Aristide march. For the first time in weeks protestors met little resistance—but this only heightened the looming sense of some final confrontation. The rage. The outrage. Thousands upon thousands were out like hornets, but there was no one upon whom to turn. "Five years!" they chanted. "Five years for Aristide!"

"That is his defined, legal term limit!" a man shouted at me. "There is a constitution! No one else will ever be our champion! No other politician will come along for us! Not if they force Ti-tide out!"

Marching along in that fuming multitude, it still seemed to me impossible that Philippe and his few hundred men could penetrate Port-au-Prince's ring of working-class neighborhoods and its riled residents. The siege of Port-au-Prince would be a protracted one. When I went to sleep that night, that's what I was thinking.

But then the dawn. The popping sound of nearby gunfire reached us well before any official confirmation that Aristide was airborne. Martin and I snuck up to the roof of the hotel, crouching behind a low wall. In the city below we could already see columns of smoke rising: from the port, along the Delmas Road, from near the palace itself. Port-au-Prince was going to immolate itself.

Soon enough we confirmed the radio-borne rumors. Aristide was indeed gone, on a plane supplied by none other than the United States and bound for who knew where. The vacuum of power would now, by necessity, be quickly filled. The problem was, we didn't know by whom. The Liberation Army, led by Guy Philippe? The Cannibal Army from Gonaives? Both groups certainly had more guns than the U.S.-supported political opposition groups in Port-au-Prince.

We grabbed our gear and drove as fast as we could down to the National Palace. Along the way we passed many houses and businesses set ablaze. Aristide supporters were looting and torching what they could. They'd also set up hundreds of impromptu roadblocks to slow any rebel advance.

A huge crowd had gathered at the palace. "I would not show yourselves," our fixer cautioned us. But there were now so many foreign journalists on the scene that we felt emboldened. We made our way to the high metal bars that surrounded the official—and now vacant—seat of power. People were crying bitterly. *Ti-tide, cinc ans! Ti-tide, cinc ans! Aristide, five years!*

Aristide militants and his ordinary Haitian supporters had long said that if Aristide were forced out they'd raze the capital. As the day wore on it appeared they were keeping their word. They gutted the port, stealing what they could there, from cars to clothes to huge chunks of frozen meat. A newswire described one looter fleeing a port warehouse wearing the front half of a horse costume.

In the industrial zone near the airport the situation was similarly chaotic. Riotous crowds smashed up factories and stole what was not nailed or soldered down. The rich Haitians who owned these sweatshops and manufacturing plants could do little to stop the pillaging, even with their well-armed cadres of bodyguards. Much of what was left of Haiti's meager industrial engine went up in smoke or disappeared on the bare backs of thieves in a matter of hours.

That afternoon President Bush ordered in the Marines to prevent the further spread of anarchy. The man who'd been steadfastly ignoring Haiti for years now told reporters that he believed it was essential that Haiti have a hopeful future.

By nightfall the Marines had landed and ensconced themselves in the National Palace. They'd met no resistance coming in. At dawn I was back before the palace gates. It was clear that the presence of U.S. troops had had an immediate and sobering effect on the Aristide mobs, who now milled about angrily on the plaza before the palace, only occasionally taunting the forward shooters from that safe distance. There would be no rioting, at least not while Marine rifles were trained on them.

As the morning wore on, the Aristide supporters must have realized there was nothing to be accomplished here and they began to go away. As word spread that the Marines were on the ground a much larger crowd of Aristide's detractors began arriving. Thousands of them clung to the bars along the perimeter of the palace, cheering *USA! USA!* The forwardmost Marines occasionally nodded, unsmiling. The rebels had still not arrived, and I wondered if they had made a strategic decision to let all of this transpire first. Clearly they would meet less resistance now. By nine o'clock Haiti's wealthy families began showing up in their Ray-Ban sunglasses and expensive four-by-fours. A truck decorated for that year's canceled carnival celebrations rolled into view blaring *rara* music. Clearly the party was just beginning.

Then a cavalcade of old cars and pickups zoomed down the wide road in front of the palace, honking like it was a wedding. On the roofs and in the beds the most ragtag looking group of grinning men hoisted their rifles into the air. Some wore ill-fitting army helmets; others wore ski hats. The rebels had arrived—Liberation and Cannibal—and their timing was impeccable. Too much so to think it hadn't been planned.

They made sure not to drive too close to the palace itself because the Marines on the perimeter had raised their own rifles. Philippe and his men descended from their vehicles and after a bit of parading back and forth they ran to the far side of the plaza and installed themselves in a nearby police station building. As they entered the premises the crowds surged toward them, cheering and singing. In his enthusiasm to join the rebels, one man tried to scale the fence around the station. A rebel who must have deemed him overly zealous shot him dead to the ground. The Marines across the plaza stirred at the sound of gunfire.

Now there commenced long uncertain minutes in which neither the rebels nor the Marines seemed to be doing much of anything but watching each other. You would not have called it a standoff because the rebels were not idiots. Still, everyone was just waiting on the plaza to see what both sides would do next. The crowd grew and grew. Tens of thousands of jubilant people were circulating continuously, running past the palace, then the rebel base, hands raised high in exultation. It occurred to me that we were living out one of those rare and ephemeral moments of uncomplicated hope in Haiti. A fleeting instant between the end of one regime and the installation of another. An uncontaminated gap in which anything was possible, anything could be dreamed.

Into this euphoria there suddenly rolled a second caravan, this time of colorful Haitian pickup trucks and small buses. The vehicles drove to the center of the plaza, also honking, and stopped. From the lead bus emerged the last person I had expected to see . . .

Reggie Barnes.

He'd taken off his jacket and rolled up his sleeves and put on sunglasses. He climbed to the roof of the bus like one of the Blues Brothers, waving to the impassive Marines. Then he knelt and extended one arm back down toward the bus's back doors. Out came a slight Indian man with a mustache wearing a Nehru jacket too warm for this place. This had to be the famous K. A. Paul.

With Reggie's assistance the preacher clambered to the roof and stood up.

Dr. Paul's mystery caravan had drawn that entire multitude around it. Once on his feet, the doctor raised his palms to the sky like Christ the Redeemer over Rio. The crowd went unexpectedly silent. The Marines and rebels both appeared fixated by this oddly dressed figure and his unknown agenda.

"Close your eyes!" Dr. Paul shouted to the people and especially, I noticed, to the several TV cameras now filming him. "Close your eyes, my children!" he yelled. "And pray to Jesus with me!" He closed his own eyes and raised his face toward the sky and was saying something. But quickly enough you could no longer hear the prayer because someone below had noticed that Dr. Paul's bus was not empty. It, and the other trucks, were all jam-packed with food. The hungry people began mobbing the vehicles, pushing and rocking them in the process. Dr. Paul was nearly thrown to the ground, but Reggie managed to catch him and help him climb back down.

"Order please! Order here!" Dr. Paul shouted. "Let's have some order here!" When he saw that no one was the slightest bit interested in order but rather in eating, he began playing catch-up, dispensing the food and other supplies to the crowd as fast as he could. He was smiling nervously and throwing items into the air like Frisbees. I wondered where he'd gotten these goods, if perhaps they weren't left over from some other mission in a country with very different necessities. Boxes of frozen pizzas, kitty litter, lubricating jelly, and diapers were soon crisscrossing the sky. When each item landed it had the effect of sparking an isolated scuffle among the desperate recipients nearest to it. They tussled like fans fighting over a home-run baseball—or worse. One man near me lurched past and into Dr. Paul's bus, snagging a bag of potatoes. But as he tried to run the sack caught on something and ripped open and the contents spilled out on the street. Young men began diving to the asphalt. Some were trampled and stood up to punch the nearest person in compensation. The individual scraps now extended and intertwined and became one indistinguishable brawl. K. A. Paul was sweating and shouting something about God and forming a single line. Finally he was crying simply "Stop! Stop!" But not one person was paying attention to him. The food riot and the fighting lasted a full twenty minutes. The

Marines watched the whole thing with detached curiosity. The rebels no longer seemed to be paying attention at all.

Eventually the caravan had been stripped bare and Reggie was able to get a rather flummoxed Dr. Paul into the front seat of the forward van. As Reggie himself was climbing in to drive away, people continued to claw at him. Nobody wanted to believe the food was entirely gone. One man approached and began shouting "Sir! Sir! A word!"

Reggie stopped. "What is it?" he barked.

"I want to tell you something, sir," the man said in English, trembling with rage. "What you've done here is shameful. You cannot come to Haiti where people are starving to death and just throw food at us like dogs. There are proper channels to distribute food. Shame on you for treating us like animals."

Reggie looked at him, then at the street littered with crushed TV dinners thawing on the asphalt and squashed fruit that no one could now eat.

"Go tell it to your ex-president," he spit. "And while you're at it tell him to get a broom."

With that final swipe, the latest peace initiative of Dr. K. A. Paul came to an end. The greatest humanist on earth was driven away in his rickety caravan and not seen again.

March 10, 2004. Mexico City. Anne had returned home to the house on Jojutla Street and was waiting for me to return. In his nearby residence President Vicente Fox was waiting hopefully for a guest-worker bill that the U.S. Congress would never approve. In Rio de Janeiro Martin Kaste was home now too, reunited with Amy. In the Central African Republic, Jean-Bertrand Aristide, a virtual prisoner, was telling the world that he'd been kidnapped, that he was still the rightful leader of Haiti. He beseeched his followers back home to continue their protests peacefully. In Port-au-Prince, in an operating room in the Canapé-Vert Clinic, a Spanish reporter named Ricardo Ortega lay in a pool of blood spreading to cover the entire floor. His own had mixed with that of the others wounded in an attack on celebrators marching past the National Palace minutes earlier. Ortega had been shot in the thorax by either a Haitian or a U.S. Marine, and he repeated over and over again in three languages as he died so that those over-hearing might get it right: *mi nombre es ricardo ortega soy espanol soy periodista je m'appelle ricardo ortega je suis espagnol je suis journaliste my name is ricardo ortega i am spanish i am a journalist* . . . Outside in the clinic courtyard his girlfriend spun like a slow statue on some senseless, rotating pedestal, her hands covering her mouth, waiting for someone to hold her. She had not yet begun to cry.

And down the hill, by the National Palace, Aristide's foes and friends continued to clash near the scene of the attack as the Marines drove about powerlessly in big green personnel carriers.

I had seen all of this, and now I was rolling toward the Port-au-Prince morgue again. The morgue is a place you must visit from time to time when each side in an armed insurrection claims it is being butchered by the other. Most casualties, at least urban ones, eventually get collected off streets and taken to hospitals, and from there they are sent to the public freezers. The problem today was that those freezers had stopped working. There was no power in this part of downtown and there hadn't been for days. The morgue had become like any other building in the tropics without ai-conditioning.

We went in through the front door, my fixer and I, and stood in a waiting area lit only by sunlight through one window. Here, where in normal times physicians received hospital workers, paramedics, police, and families, the smell was already like a liquid.

A short, slight orderly in a gray smock received us without question and opened a further door that led down a very dark hallway. I willed myself to inhale through my mouth alone, and so far, after that first mistaken breath, it was working. We walked down the hall and the door shut behind us, making it nearly impossible to see at first. Eventually we reached a large room used as an intake or processing point. From here cadavers went either into an autopsy room or into the freezers until the living deliberated their end. Our eyes slowly adjusted to the dimness.

"Who are they?" I asked.

"Which ones?" said the orderly.

"The two with their hands tied," I said.

"I don't know their names, sir. They had no identification when they were brought in. Virtually none do."

The two young men both had their hands bound behind their backs with rope and they appeared to have been pulled some distance behind vehicles. Their clothes were ripped nearly away and then the skin, especially on their scalps. I could see that they'd just been dragged in here because there were dark wet slide marks leading to their bodies.

"Who brought them?"

"Men. It is hoped that their families will come. That they will know to come here to look for them."

"Will they stay there, on the floor?"

"There is no other place, sir."

With that the orderly approached a huge metal freezer door, the kind you might see in the kitchen of a big hotel.

"Shall I open it?" he whispered.

I hadn't expected quite this.

No one said anything. The orderly pulled down on the steel latch like the handle of an old-fashioned water pump. There was a metallic clunk and a sigh as some seal broke. I moved out of the way as the heavy door swung open. Then I stepped forward and in the near dark I raised my camera, took approximate aim, and pressed the shutter button. The flash fired once, twice, five times in succession. In those bursts of light you could at last understand the rituals around last rites and burial. Here was the absence of those attentions. The bodies were stacked four layers high. It was safe to say that some had been here for more than a week. The last several entries had been tossed in with less attention to detail because the pile had grown too large to manage. The young and the old half-slid from shelves, some partially sitting. Shoeless feet poking toward the ceiling. Everyone's last outfits mussed or partly missing. You could process the sight of it, and even the smell, but what you couldn't make sense of was the silence. I had never seen so many bodies, and somehow I was expecting some complaint. Some outright, raucous complaint. But the only one here to protest this affront was the orderly, and he was closing the door again. Perhaps for him guarding silence was akin to respect. Either way, what could he say and to whom? I was not thinking straight.

I stepped backward and bumped up against a metal, wheeled gurney that I had not seen. When I turned to move around it my fingers brushed against a tiny hand cold as a trout. There were two on the gurney, both girls, neither more than a year old. Dressed neatly in warm clothes and caps that their aggrieved must have thought would serve them here, at least symbolically. They were the size of perfect dolls and might otherwise have been sleeping.

I pointed.

"They were not killed in fighting," the orderly said softly. "But they might not have died." He stopped talking and waved his arms. They might not have died, he meant to say, but they did because there were no doctors to see them. The fact that the orderly had not thrown them in the freezer struck me as a gesture of tenderness. Or maybe he just hadn't found the time yet.

In my head there was a sound like a cork popping. I stumbled back down the long hallway, holding to the back of my fixer's shirt. When we hit the fresh air he doubled over and vomited onto high weeds along a wall. I sucked in air through my mouth as hard as I could and blew out through my nose several times before taking a normal breath. But our clothes were saturated with it. It had been impossible to tell who the dead were or even how they had died. I wanted to go back inside and save those baby girls from that. To bury them out here, somewhere, anywhere, in the light. Then my phone rang. It was Anne.

"*Oye*," she asked down the crackling line, "how do you say 'baby girl' in Kreyol?"

"What the *hell?*" I yelled. "Have you lost your mind! What in the *hell* would cause you to say that?"

There was a moment of silence.

"Are you okay?" she asked, half worried, half accusing. She'd come to believe that I was hiding the worst from her. There was a continual disconnect between my version of events and what she was seeing from home. "What's going on over there?"

"Nothing, nothing," I said, searching for steadiness. "Everything's fine. It's all over now."

"What is over?"

"Anne, please, what the devil are you talking about?"

"I was talking about our baby," she said, and I could hear the joy suppressed in her voice now. "We're going to have a baby girl."

EPILOGUE

When the U.S. Marines and an army of UN bureaucrats had gotten a momentary handle on Haiti's unrest I was able to fly back to Mexico City. More than a month had passed. The first thing I did was put my ear to the slight, warm swell of Anne's belly. Then I quit my job.

The worst of the Haitian crisis was behind us, but there was going to be a next time. And a next. And I didn't want to hang around for them. Choosing over and over between my future family, whatever our circumstances, and the story, whatever its weight. I could not imagine such a life now. At least not a life with others. To do this job takes a certain type of person, in a certain period of his or her life. I'd come down here fitting the profile, but with time my profile had changed. Before lunch on that first day back in Mexico I emailed in my resignation. It was sad to interrupt a dream only partly realized. But it felt right and made sense because there was this other one just beginning.

July 27, 2004. I sat down behind my desk on Jojutla Street 13 for the last time. Four floors of accumulated belongings were now in packing boxes. Downstairs, movers had just loaded them all into trucks. Downstairs, Anne was on the phone calling friends to take us to the airport.

This was it. The only items left on my desk: the phone and a red plastic bag of raisins that I'd found in a final sweep through the kitchen. I figured I should eat them rather than leave them behind for NPR's next reporter. Who knew how long it would take the network to replace me?

Out the open westerly window the Chapultepec Castle simmered darkly under the afternoon sun. How many times had I imagined a boy wrapped in a flag leaping from it. I looked around the office, at the shelves filled with all those books. Biography, science fiction, philosophy. Bertrand Russell's *Problems of Philosophy* was long back in its proper spot and had grown dusty again. Who would ever know its story? The door to the guest room stood open, and there too I could see no evidence of all the comings and goings, no signs of what guests and refugees had slept there. Of what traces we'd left upon each other. I thought, how do you say good-bye to the bygone?

And who should return with an answer? As I sat in my chair with my two hands planted on the rough pine desktop, knowing that things could not be otherwise, the bag of raisins began to move before my eyes. It slid very slowly and was rotating, so slowly that my first thought was of the wind entering through the window. Then I looked to see if the phone cord or some other thing might be affecting it. I peered from above, considering even some raisin avalanche within the bag itself. Nothing checked out—yet there was the bag in motion. The episode, or dare I say gesture, lasted a full twenty seconds.

When it was over I waited a long moment for more. And when nothing more happened I said out loud, "Is that all you've got?" I slapped my hands on the desk and just grinned. "You guys wrote the book on scary," I said, "and this is your parting performance. You twirl a bag of raisins."

But I was getting it all wrong. I was the one who was departing, not them. And anyway, what did I expect: A chorus line of skeletons? A big hug from Pants Boy? It had been more than a year since I'd paid the ghosts any attention, even longer since the last time I'd asked them explicitly to stay, hardly meaning it but trying so hard to. That had been my first struggle in this house. I'd never made much headway in my own heart, but at least there'd been the effort. And then, with the onset of so many other struggles, the effort had dried up. I'd even tried to paint over them, to interior-decorate them away. After that the ghosts grew quiet. Until at some point I believed they were gone and I forgot about them.

Now here they were again, or here he was, or she: I didn't

even know. Had they ever left at all, whatever they were?
What matter that you can't see a thing with your eyes. We
continue, their point seemed to be now, whether or not
you have faith in our perpetuity. Maybe this was what they
wanted me to understand. I liked the idea because, seen this
way, there were no good-byes. All the people you'd met and
every drop of sweat and blood were memories yet intact.
They were the ghosts that did not leave you.

Or maybe this was not the point. Maybe there was none
at all and there never had been. As intriguing as closet doors
bursting open and mobile bags of fruit might be, maybe they
were as meaningless as the electrical twitch of a dead lab
frog's leg.

I was not going to get to the bottom of it, but that was
in keeping with this life now ending. I'd long ago reached
accommodation in my house without answers, and there
was no reason to expect any now. But just to be safe, just
in case we were communicating, just in case my neglected
Buddhist masters were right about all this suffering, I stood
up and said, "Well, welcome. As always, welcome to you all."
I nodded to each silent corner of the room, feeling as silly
as I had on that first night when I'd addressed this invisible audience. "You're more than welcome to stay," I said,
"every last one of you. Hell, there's more room here than I
ever knew what to do with anyway." I picked up the bag of
raisins. Then something else occurred to me. A last-ditch
message to boost their morale and maybe even leave behind
a trace of my own.

"Assuming you do decide to stick around," I said, "feel free to start spooking people again. My replacement should be along soon enough."

I tried with all of my heart to want them to stay, and this time was the easiest. Because I was out of there. I offered out what was left of the raisins like birdseed onto the desk, then went downstairs to catch my flight out with Anne and our unborn child.

ACKNOWLEDGMENTS

The author would like to gratefully thank Dzongsar Khyentse Rinpoche, Chogyam Trungpa Rinpoche, Trish and Sandy Hadden, Sarah Burnes, Michael Signorelli, the meat-cleaver-wielding Bernadette Rivero, Larry Collins, Sloan Harris, Michael Deibert, Marion Lloyd, Lindsey Addario, Franc Contreras, Martin Kaste, David Gonzalez, Erin Hennessey, Bob and Francine Cassuto, Karla Bluntschli, Chantal Regnault, Mario Delatour, Gunnar Vatvedt, Elena Vialo, and Sigur Rós, among many others—including, it would seem, Lawrence Collins and Daniel Sullivan.

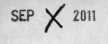